Mountain Laurel Waldorf School
P.O. Box 488
New Paltz, N.Y. 12561

Onteora

Challenging Problems in Algebra

ALFRED S. POSAMENTIER

Professor of Mathematics Education
The City College of the City University of New York

CHARLES T. SALKIND

Late Professor of Mathematics
Polytechnic University, New York

DOVER PUBLICATIONS, INC.
New York

Published in Canada by General Publishing Company, Ltd., 30 Lesmill Road, Don Mills, Toronto, Ontario.
Published in the United Kingdom by Constable and Company, Ltd., 3 The Lanchesters, 162–164 Fulham Palace Road, London W6 9ER.

Bibliographical Note

This Dover edition, first published in 1996, is an unabridged, very slightly altered republication of the work first published in 1970 by the Macmillan Company, New York, and again in 1988 by Dale Seymour Publications, Palo Alto, California. For the Dover edition, Professor Posamentier has made two slight alterations in the introductory material.

Library of Congress Cataloging-in-Publication Data

Posamentier, Alfred S.
　　Challenging problems in algebra / Alfred S. Posamentier, Charles T. Salkind.
　　　　p.　　cm.
　　"An unabridged, very slightly altered republication of the work first published in 1970 by the Macmillan Company, New York, and again in 1988 by Dale Seymour Publications, Palo Alto, California. For the Dover edition, Professor Posamentier has made two slight alterations in the introductory material"
—T.p. verso.
　　ISBN 0-486-69148-9 (pbk.)
　　1. Algebra—Problems, exercises, etc. I. Salkind, Charles T., 1898– . II. Title.
QA157.P68　1996
512.9'076—dc20　　　　　　　　　　　　　　　　　　　　　95-48215
　　　　　　　　　　　　　　　　　　　　　　　　　　　　　　　CIP

Manufactured in the United States of America
Dover Publications, Inc., 31 East 2nd Street, Mineola, N.Y. 11501

CONTENTS

INTRODUCTION

The challenge of well-posed problems transcends national boundaries, ethnic origins, political systems, economic doctrines, and religious beliefs; the appeal is almost universal. Why? You are invited to formulate your own explanation. We simply accept the observation and exploit it here for entertainment and enrichment.

This book is a new, combined edition of two volumes first published in 1970. It contains more than three hundred problems that are "off the beaten path"—problems that give a new twist to familiar topics or that introduce unfamiliar topics. With few exceptions, their solution requires little more than some knowledge of elementary algebra, though a dash of ingenuity may help. The problems range from fairly easy to hard, and many have extensions or variations that we call *challenges*. Supplied with pencil and paper and fortified with a diligent attitude, you can make this material the starting point for exploring unfamiliar or little-known aspects of mathematics. The challenges will spur you on; perhaps you can even supply your own challenges in some cases. A study of these non-routine problems can provide valuable underpinnings for work in more advanced mathematics.

This book, with slight modifications made, is as appropriate now as it was a quarter century ago when it was first published. The National Council of Teachers of Mathematics (NCTM), in their *Curriculum and Evaluation Standards for High School Mathematics* (1989), lists problem solving as its first standard, stating that "mathematical problem solving in its broadest sense is nearly synonymous with doing mathematics." They go on to say, "[problem solving] is a process by which the fabric of mathematics is identified in later standards as both constructive and reinforced."

This strong emphasis on mathematics is by no means a new agenda item. In 1980, the NCTM published *An Agenda for Action*. There, the NCTM also had problem solving as its first item, stating, "educators should give priority to the identification and analysis of specific problem solving strategies .... [and] should develop and disseminate examples of 'good problems' and strategies." It is our intention to provide secondary mathematics educators with materials to help them implement this very important recommendation.

ABOUT THE BOOK

Challenging Problems in Algebra is organized into three main parts: "Problems," "Solutions," and "Answers." Unlike many contemporary problem-solving resources, this book is arranged *not* by problem-solving technique, but by topic. We feel that announcing the technique to be used stifles creativity and destroys a good part of the fun of problem solving.

The problems themselves are grouped into two sections. Section I covers eight topics that roughly parallel the sequence of a first year algebra course. Section II presents twelve topics that roughly parallel the second year algebra course.

Within each topic, the problems are arranged in approximate order of difficulty. For some problems, the basic difficulty may lie in making the distinction between relevant and irrelevant data or between known and unknown information. The sure ability to make these distinctions is part of the process of problem solving, and each devotee must develop this power by him- or herself. It will come with sustained effort.

In the "Solutions" part of the book, each problem is restated and then its solution is given. From time to time we give alternate methods of solution, for there is rarely only one way to solve a problem. The solutions shown are far from exhaustive, and intentionally so, allowing you to try a variety of different approaches. Particularly enlightening is the strategy of using multiple methods, integrating algebra, geometry, and trigonometry. Instances of multiple methods or multiple interpretations appear in the solutions. Our continuing challenge to you, the reader, is to find a different method of solution for every problem.

The third part of the book, "Answers," has a double purpose. It contains the answers to all problems and challenges, providing a quick check when you have arrived at a solution. Without giving away the entire solution, these answers can also give you a hint about how to proceed when you are stuck.

Appendices at the end of the book provide information about several specialized topics that are not usually covered in the regular curriculum but are occasionally referred to in the solutions. This material should be of particular interest and merits special attention at the appropriate time.

THE TOPICS COVERED

Section I. The book begins with a chapter of general problems, simple to state and understand, that are generally appealing to students. These should serve as a pleasant introduction to problem solving early in the elementary algebra course.

Chapter 2 demonstrates the true value of algebra in understanding arithmetic phenomena. With the use of algebraic methods, students are guided through fascinating investigations of arithmetic curiosities.

Familiar and unfamiliar relations are the bases for some cute problems in chapter 3. A refreshing consideration of various base systems is offered in chapter 4. Uncommon problems dealing with the common topics of equations, inequalities, functions, and simultaneous equations and inequalities are presented along with stimulating challenges in chapters 5, 6, and 7.

The last chapter of this section contains a collection of problems summarizing the techniques encountered earlier. These problems are best saved for the end of the elementary algebra course.

Section II. The second section opens with a chapter on one of the oldest forms of algebra, Diophantine equations—indeterminate equations for which only integer solutions are sought. These problems often appear formidable to the young algebra student, yet they can be solved easily after some experience with the type (which this section offers).

The next two chapters present some variations on familiar themes, functions and inequalities, treated here in a more sophisticated manner than was employed in the first section.

The field of number theory includes some interesting topics for the secondary school student, but all too often this area of study is avoided. Chapter 12 presents some of these concepts through a collection of unusual problems. Naturally, an algebraic approach is used throughout.

Aside from a brief exposure to maxium and minimum points on a parabola, very little is done with these concepts prior to a study of the calculus. Chapter 13 will demonstrate through problem solving some explorations of these concepts at a relatively elementary level.

Chapters 14, 15, 17, and 18 offer unconventional problems for some standard topics: quadratic equations, simultaneous equations, series, and logarithms. The topic of logarithms is presented in this book as an end in itself rather than as a (computational) means to an end, which has been its usual role. Problems in these chapters should shed some new (and dare we say refreshing) light on these familiar topics.

Chapter 16 attempts to bring some new life and meaning, via problem solving, to analytic geometry. Chapter 19 should serve as a motivator for further study of probability and a consideration of general counting techniques.

We conclude our treatment of problem solving in algebra with chapter 20, "An Algebraic Potpourri." Here we attempt to pull together some of the problems and solution techniques considered in earlier sections. These final problems are quite challenging as well as out of the ordinary, even though the topics from which they are drawn are quite familiar.

USING THE BOOK

This book may be used in a variety of ways. It is a valuable supplement to the basic algebra textbooks, both for further explorations on specific topics and for practice in developing problem-solving techniques. The book also has a natural place in preparing individuals or student teams for participation in mathematics contests. Mathematics clubs might use this book as a source of independent projects or activities. Whatever the use, experience has shown that these problems motivate people of all ages to pursue more vigorously the study of mathematics.

Very near the completion of the first phase of this project, the passing of Professor Charles T. Salkind grieved the many who knew and respected him. He dedicated much of his life to the study of problem posing and problem solving and to projects aimed at making problem solving meaningful, interesting, and instructive to mathematics students at all levels. His efforts were praised by all. Working closely with this truly great man was a fascinating and pleasurable experience.

Alfred S. Posamentier
1996

PREPARING TO
SOLVE A PROBLEM

A strategy for attacking a problem is frequently dictated by the use of analogy. In fact, searching for an analogue appears to be a psychological necessity. However, some analogues are more apparent than real, so analogies should be scrutinized with care. Allied to analogy is structural similarity or pattern. Identifying a pattern in apparently unrelated problems is not a common achievement, but when done successfully it brings immense satisfaction.

Failure to solve a problem is sometimes the result of fixed habits of thought, that is, inflexible approaches. When familiar approaches prove fruitless, be prepared to alter the line of attack. A flexible attitude may help you to avoid needless frustration.

Here are three ways to make a problem yield dividends:

(1) The result of formal manipulation, that is, "the answer," may or may not be meaningful; find out! Investigate the possibility that the answer is not unique. If more than one answer is obtained, decide on the acceptabiklity of each alternatibe. Where appropriate, estimate the answer in advance of the solution. The habit of estimating in advance should help to prevent crude errors in manipulation.

(2) Check possible restrictions on the data and/or the results. Vary the data in significant ways and study the effect of such variations on the original result.

(3) The insight needed to solve a generalized problem is sometimes gained by first specializing it. Conversely, a specialized problem, difficult when tackled directly, sometimes yields to an easy solution by first generalizing it.

As is often true, there may be more than one way to solve a problem. There is usually what we will refer to as the "peasant's way" in contrast to the "poet's way"—the latter being the more elegant method.

To better understand this distinction, let us consider the following problem:

> If the sum of two numbers is 2, and the product of these
> same two numbers is 3, find the sum of the reciprocals
> of these two numbers.

Those attempting to solve the following pair of equations simultaneously are embarking on the "peasant's way" to solve this problem.

$$x + y = 2$$
$$xy = 3$$

Substituting for y in the second equation yields the quadratic equation, $x^2 - 2x + 3 = 0$. Using the quadratic formula we can find $x = 1 \pm i \sqrt{2}$. By adding the reciprocals of these two values of x, the answer $\frac{2}{3}$ appears.

This is clearly a rather laborious procedure, not particularly elegant.

The "poet's way" involves working backwards. By considering the desired result

$$\frac{1}{x} + \frac{1}{y}$$

and seeking an expression from which this sum may be derived, one should inspect the algebraic sum:

$$\frac{x + y}{xy}$$

The answer to the original problem is now obvious! That is, since $x + y = 2$ and $xy = 3$, $\frac{x + y}{xy} = \frac{2}{3}$. This is clearly a more elegant solution than the first one.

The "poet's way" solution to this problem points out a very useful and all too often neglected method of solution. A reverse strategy is certainly not new. It was considered by Pappus of Alexandria about 320 A.D. In Book VII of Pappus' *Collection* there is a rather complete description of the methods of "analysis" and "synthesis." T. L. Heath, in his book *A Manual of Greek Mathematics* (Oxford University Press, 1931, pp. 452-53), provides a translation of Pappus' definitions of these terms:

> *Analysis* takes that which is sought as if it were admitted and passes from it through its successive consequences to something which is admitted as the result of synthesis: for in analysis we assume that which is sought as if it were already done, and we inquire what it is from which this results, and again what is the antecedent cause of the latter, and so on, until, by so retracing our steps, we come upon something already known or belonging to the class of first principles, and such a method we call analysis as being solution backward.

> But in *synthesis*, reversing the progress, we take as already done that which was last arrived at in the analysis and, by arranging in their natural order as consequences what before were antecedents, and successively connecting them one with another, we arrive finally at the construction of that which was sought: and this we call *synthesis*.

Unfortunately, this method has not received its due emphasis in the mathematics classroom. We hope that the strategy recalled here will serve you well in solving some of the problems presented in this book.

Naturally, there are numerous other clever problem-solving strategies to pick from. In recent years a plethora of books describing various problem-solving methods have become available. A concise description of these problem-solving strategies can be found in *Teaching Secondary School Mathematics: Techniques and Enrichment Units,* by A. S. Posamentier and J. Stepelman, 4th edition (Columbus, Ohio: Prentice Hall/Merrill, 1995).

Our aim in this book is to strengthen the reader's problem-solving skills through nonroutine motivational examples. We therefore allow the reader the fun of finding the best path to a problem's solution, an achievement generating the most pleasure in mathematics.

PROBLEMS

SECTION I
First Year Algebra

1. Posers: Innocent and Sophisticated

The number of hairs on a human head, a castaway on a desert island trying to conserve his supply of water, a stubborn watch that stops for fifteen minutes every hour — these are some of the fanciful settings for the problems in this opening section. A variety of mathematical ideas are encountered in the problems, with the Pigeon Hole Principle and the mathematics of uniform motion receiving the greatest share of attention.

1-1 Suppose there are 6 pairs of blue socks all alike, and 6 pairs of black socks all alike, scrambled in a drawer. How many socks must be drawn out, all at once (in the dark), to be certain of getting a matching pair?

Challenge 1 Suppose the drawer contains 3 black pairs of socks, 7 green pairs, and 4 blue pairs, scrambled. How many socks must be drawn out, all at once (in the dark), to be certain of getting a matching pair?

Challenge 2 Suppose there are 6 different pairs of cuff links scrambled in a box. How many links must be drawn out, all at once (in the dark), to be certain of getting a matching pair?

1-2 Find five positive whole numbers a, b, c, d, e such that there is no subset with a sum divisible by 5.

1-3 A multiple dwelling has 50 letter boxes. If 101 pieces of mail are correctly delivered and boxed, show that there is at least one letter box with 3 or more pieces of mail.

Challenge 1   What conclusion follows if there are 102 pieces of mail?

Challenge 2   What conclusion follows if there are 150 pieces of mail?

Challenge 3   What conclusion follows if there are 151 pieces of mail?

Challenge 4   If no human being has more than 300,000 hairs on his head, and the population of Megalopolis is 8,000,000 persons, what is the least value of n in the assertion that there are n persons in Megalopolis with the same number of hairs on their heads? (Assume that all people have at least one hair on their head.)

1-4 Assume that at least one of a_1 and b_1 has property P, and at least one of a_2 and b_2 has property P, and at least one of a_3 and b_3 has property P. Prove that at least two of a_1, a_2, a_3, or at least two of b_1, b_2, b_3 have property P.

Challenge 1   Add to the information in Problem 1-4 that at least one of a_4 and b_4 has property P, and that at least one of a_5 and b_5 has property P. Prove that at least 3 of the a's, or at least 3 of the b's have property P.

Challenge 2   Assume that property Q is possessed by at least one of a_1, b_1, c_1, by at least one of a_2, b_2, c_2, ..., by at least one of a_{10}, b_{10}, c_{10}. Find the largest value of k in the assertion that at least k of the a's, or at least k of the b's, or at least k of the c's have property Q.

Challenge 3   Assume that property R is possessed by at least two of a_1, b_1, c_1, by at least two of a_2, b_2, c_2, ..., by at least two of a_5, b_5, c_5. Find the largest value of m for which it can be said that at least m of the a's, or of the b's, or of the c's have property R.

1-5 An airplane flies round trip a distance of L miles each way. The velocity with head wind is 160 m.p.h., while the velocity with tail wind is 240 m.p.h. What is the average speed for the round trip?

1-6 Assume that the trains between New York and Washington leave each city every hour on the hour. On its run from Washington to New York, a train will meet n trains going in the opposite direction. If the one-way trip in either direction requires four hours exactly, what is the value of n?

Challenge   Change "four hours exactly" to "three and one-half hours exactly" and solve the problem.

1-7   A freight train one mile long is traveling at a steady speed of 20 miles per hour. It enters a tunnel one mile long at 1 P.M. At what time does the rear of the train emerge from the tunnel?

1-8   A watch is stopped for 15 minutes every hour on the hour. How many actual hours elapse during the interval the watch shows 12 noon to 12 midnight?

Challenge 1   A watch is stopped for 15 minutes every hour on the hour. According to this watch, how many hours elapse between 12 noon and 12 midnight (actual time)?

Challenge 2   Between 12 noon and 12 midnight, a watch is stopped for 1 minute at the end of the first full hour, for 2 minutes at the end of the second full hour, for 3 minutes at the end of the third full hour, and so forth for the remaining full hours. What is the true time when this watch shows 12 midnight?

1-9   The last three digits of a number N are $x25$. For how many values of x can N be the square of an integer?

1-10   A man born in the eighteenth century was x years old in the year x^2. How old was he in 1776? (Make no correction for calendric changes.)

Challenge 1   Is there a corresponding puzzle for the nineteenth century? If so, find the man's age in 1876.

Challenge 2   Show that there is no corresponding puzzle for the twentieth century.

1-11   To conserve the contents of a 16 oz. bottle of tonic, a castaway adopts the following procedure. On the first day he drinks 1 oz. of tonic and then refills the bottle with water; on the second day he drinks 2 oz. of the mixture and then refills the bottle with water; on the third day he drinks 3 oz. of the mixture and again refills the bottle with water. The procedure is continued for succeeding days until the bottle is empty. How many ounces of water does he thus drink?

Challenge Assume that the castaway drinks only $\frac{1}{2}$ oz. the first day, 1 oz. the second day, $1\frac{1}{2}$ oz. the third day, and so forth. Find the total consumption of water.

1-12 Which yields a larger amount with the same starting salary:
Plan I, with four annual increases of $100 each, or
Plan II, with two biennial increases of $200 each?

Challenge How does the result change if the increase under Plan II is $250?

1-13 Assuming that in a group of n people any acquaintances are mutual, prove that there are two persons with the same number of acquaintances.

1-14 The smallest of n consecutive integers is j. Represent in terms of j (a) the largest integer L (b) the middle integer M.

Challenge 1 Let j be the largest of n consecutive integers. Represent in terms of j (a) the smallest integer S (b) the middle integer M.

Challenge 2 Let j be the smallest of n consecutive even integers. Represent in terms of j (a) the largest integer L (b) the middle integer M.

Challenge 3 Let j be the smallest of n consecutive odd integers. Represent in terms of j (a) the largest integer L (b) the middle integer M.

Challenge 4 If n is a multiple of 4, find the integer in position $\frac{3n}{4}$ for the original problem and each of Challenges 1, 2, and 3.

1-15 We define the symbol $|x|$ to mean the value x if $x \geq 0$, and the value $-x$ if $x < 0$. Express $|x - y|$ in terms of $\max(x, y)$ and $\min(x, y)$ where $\max(x, y)$ means x if $x > y$, and y if $x < y$, and $\min(x, y)$ means x if $x < y$, and y if $x > y$.

Challenge 1 Does the result cover the case when $x = y$?

Challenge 2 Prove that $\max(x, y) = \frac{x + y}{2} + \frac{|x - y|}{2}$, and find the corresponding expression for $\min(x, y)$.

1-16 Let $x^+ = \begin{cases} x \text{ if } x \geq 0 \\ 0 \text{ if } x < 0, \end{cases}$ and let $x^- = \begin{cases} -x \text{ if } x \leq 0 \\ 0 \text{ if } x > 0. \end{cases}$

Express:

(a) x in terms of x^+ and x^- (b) $|x|$ in terms of x^+ and x^-

(c) x^+ in terms of $|x|$ and x (d) x^- in terms of $|x|$ and x.

(See problem 1-15 for the meaning of $|x|$.)

1-17 We define the symbol $[x]$ to mean the greatest integer which is not greater than x itself. Find the value of $[y] + [1 - y]$.

Challenge 1 Find the value of (a) $[y] - [1 - y]$ (b) $[1 - y] - [y]$.

Challenge 2 Evaluate $F = \dfrac{[x]}{x}$ when (a) x is an integer ($x \neq 0$) (b) x is a positive non-integer (c) x is a negative non-integer.

Challenge 3 Evaluate $D = [x^2] - [x]^2$ when (a) $0 < x < 1$ (b) $1 < x < 2$ (c) $2 < x < 3$.

Challenge 4 Find an x satisfying the equation $[x]x = 11$.

Challenge 5 Let $(x) = x - [x]$; express $(x + y)$ in terms of (x) and (y).

1-18 At what time after 4:00 will the minute hand overtake the hour hand?

Challenge 1 At what time after 7:30 will the hands of a clock be perpendicular?

Challenge 2 Between 3:00 and 4:00 Noreen looked at her watch and noticed that the minute hand was between 5 and 6. Later, Noreen looked again and noticed that the hour hand and the minute hand had exchanged places. What time was it in the second case?

Challenge 3 The hands of Ernie's clock overlap exactly every 65 minutes. If, according to Ernie's clock, he begins working at 9 A.M. and finishes at 5 P.M., how long does Ernie work, according to an accurate clock?

---

2   Arithmetic: Mean and Otherwise

"Mathematics is the queen of the sciences, and arithmetic is her crown," said the great mathematician Carl Friedrich Gauss. School arithmetic eventually grows up and turns into the branch of mathematics called number theory, which has fascinated mathematicians and amateurs alike through the ages. In this section, you will find problems from number theory involving such topics as means, factorization, primes, divisibility, partitions, and remainders.

2-1 The arithmetic mean (A.M.), or ordinary average, of a set of 50 numbers is 32. The A.M. of a second set of 70 numbers is 53. Find the A.M. of the numbers in the sets combined.

Challenge 1   Change the A.M. of the second set to -53, and solve.

Challenge 2   Change the number of elements in each set to 1, and solve.

Challenge 3   Find the point-average of a student with A in mathematics, A in physics, B in chemistry, B in English, and C in history — using the scale: A, 5 points; B, 4 points; C, 3 points; D, 1 point — when (a) the credits for the courses are equal (b) the credits for the courses are mathematics, 4; physics, 4; chemistry, 3; English, 3; and history, 3.

Challenge 4   (a) Given n numbers each equal to $1 + \frac{1}{n}$, and two numbers each equal to 1; find their A.M. (b) Given n numbers each equal to $1 + \frac{1}{n}$, and one number 1; find their A.M. Which of (a) and (b) is larger?

Challenge 5   Given n numbers each equal to $1 - \frac{1}{n}$, and one number 2; find their A.M.

Challenge 6   In order to find the A.M. of 8 numbers $a_1, a_2, \ldots, a_8$, Carl takes one-half of $\frac{1}{4}s_1 + \frac{1}{4}s_2$ where $s_1 = a_1 + a_2 + a_3 + a_4$, and $s_2 = a_5 + a_6 + a_7 + a_8$; and Caroline takes one-half of $\frac{1}{4}s_3 + \frac{1}{4}s_4$ where $s_3 = a_1 + a_3 + $

$a_5 + a_7$, and $s_4 = a_2 + a_4 + a_6 + a_8$. Explain why both obtain the correct A.M.

Challenge 7   Estimate the approximate A.M. of the set $\{61, 62, 63, 65, 68, 73, 81, 94\}$.

2-2   Express the difference of the squares of two consecutive even integers in terms of their arithmetic mean (A.M.).

Challenge   How does the result change if two consecutive odd integers are used?

2-3   It is a fundamental theorem in arithmetic that a natural number can be factored into prime factors in only one way — if the order in which the factors are written is ignored. This is known as the Unique Factorization Theorem. For example, 12 is uniquely factored into the primes 2, 2, 3.

   Consider the set $S_1 = \{4, 7, 10, \ldots, 3k + 1, \ldots\}$, in which $k = 1, 2, \ldots, n, \ldots$. Does S_1 have unique factorization?

Challenge   Is factorization unique in $S_2 = \{3, 4, 5, \ldots, k, \ldots\}$?

2-4   What is the smallest positive value of n for which $n^2 + n + 41$ is not a prime number?

Challenge   Examine the expression $n^2 - n + 41$ for primes.

2-5   Given the positive integers a, b, c, d with $\frac{a}{b} < \frac{c}{d} < 1$; arrange in order of increasing magnitude the five quantities: $\frac{b}{a}, \frac{d}{c}, \frac{bd}{ac}, \frac{b+d}{a+c}, 1$.

2-6   It can be proved (see Appendix I) that, for any natural number n, the terminal digit of n^5 is the same as that of n itself; that is, n^5TDn, where the symbol TD means "has the same terminal digit." For example, 4^5TD4.

   Find the terminal digit of **(a)** 2^{12} **(b)** 2^{30} **(c)** 7^7 **(d)** 8^{10} **(e)** $8^{10} \cdot 7^{11}$

Challenge   Find the terminating digit of **(a)** $\left(\frac{5}{8}\right)^5$ **(b)** $\left(\frac{4}{7}\right)^5$.

2-7   If $N = 1 \cdot 2 \cdot 3 \cdots 100$ (more conveniently written 100!), find the number of terminating zeros when the multiplications are carried out.

Challenge Find the number of terminating zeros in $D = 36! - 24!$

2-8 Find the maximum value of x such that 2^x divides 21!

Challenge 1 Find the highest power of 3 in 21!

Challenge 2 Find the highest power of 2 in 21! excluding factors also divisible by 3.

2-9 The number 1234 is not divisible by 11, but the number 1243, obtained by rearranging the digits, is divisible by 11. Find all the rearrangements that are divisible by 11.

Challenge Solve the problem for 12034.

2-10 Let k be the number of positive integers that leave a remainder of 24 when divided into 4049. Find k.

Challenge 1 Find the largest integer that divides 364, 414, and 539 with the same remainder in each case.

Challenge 2 A somewhat harder problem is this: find the largest integer that divides 364, 414, and 541 with remainders R_1, R_2, and R_3, respectively, such that $R_2 = R_1 + 1$, and $R_3 = R_2 + 1$.

Challenge 3 A committee of three students, A, B, and C, meets and agrees that A report back every 10 days, B, every 12 days, and C, every 15 days. Find the least number of days before C again meets both A and B.

2-11 List all the possible remainders when an even integer square is divided by 8.

Challenge List all the possible remainders when an odd integer square is divided by 8.

2-12 Which is larger: the number of partitions of the integer $N = k \cdot 10^2$ into $2k + 1$ positive even integers, or the number of partitions of N into $2k + 1$ positive odd integers, where $k = 1, 2, 3, \ldots$? To partition a positive integer is to represent the integer as a sum of positive integers.

2-13 Given the three-digit number $N = a_1a_2a_3$, written in base 10, find the least absolute values of m_1, m_2, m_3 such that N is divisible by 7 if $m_1a_1 + m_2a_2 + m_3a_3$ is divisible by 7.

Challenge 1   Solve the problem for the six-digit number $N = a_1a_2a_3a_4a_5a_6$.
NOTE: Only $|m_1|$, $|m_2|$, and $|m_3|$ are needed.

Challenge 2   Solve the problem for the four-digit number $N = a_1a_2a_3a_4$.

2-14   When $x^3 + a$ is divided by $x + 2$, the remainder is known to be -15. Find the numerical value of a.

Challenge 1   Find the smallest value of a for which $x^3 + a$ is exactly divisible by $x + 2$.

Challenge 2   Find the value of a in the original problem when $x + 2$ is replaced by $x - 2$.

2-15   If $x - a$ is a factor of $x^2 + 2ax - 3$, find the numerical value(s) of a.

Challenge 1   Find the remainder when $P(x) = x^3 - 2x^2 + 2x - 2$ is divided by $x + 1$.

Challenge 2   Find the remainder when $x^6 + 1$ is divided by $x - m$.
Find the remainder when $x^6 + 1$ is divided by $x + m$.
Find the remainder when $x^6 + 1$ is divided by $x^2 - m$.
HINT: $x^6 + 1 = (x^2)^3 + 1$

2-16   Let N be the product of five different odd prime numbers. If N is the five-digit number $abcab$, $4 < a < 8$, find N.

2-17   If a five-digit number N is such that the sum of the digits is 29, can N be the square of an integer?

2-18   Each of the digits 2, 3, 4, 5 is used only once in writing a four-digit number. Find the number of such numbers and their sum.

2-19   Find all positive integral values of k for which $8k + 1$ expressed in base 10 exactly divides 231 expressed in base 8.

Challenge   Solve the problem with 231 expressed in base 12.

2-20   Express in terms of n the positive geometric mean of the positive divisors of the natural number n. Definition: The positive geometric mean of the k positive numbers $a_1, a_2, \ldots, a_k$ is $\sqrt[k]{a_1a_2\ldots a_k}$.

3   Relations: Familiar and Surprising

Relations defined by equations and inequalities are a common feature of the algebraic landscape. Some of the sources for the unusual relations in these problems are monetary transactions, percents, and the Cartesian lattice.

3-1 Let $y_1 = \dfrac{x+1}{x-1}$. Let y_2 be the simplified expression obtained by replacing x in y_1 by $\dfrac{x+1}{x-1}$. Let y_3 be the simplified expression obtained by replacing x in y_2 by $\dfrac{x+1}{x-1}$, and so forth.   Find y_6, y_{100}, y_{501}.

Challenge 1   Find the value of y_{200} when $x = 2$.

Challenge 2   Find the value of y_{501} when $x = 2$.

Challenge 3   Find the value of y_{201} when $x = 1$. Be careful!

Challenge 4   Find the value of y_{200} when $x = 1$. Be doubly careful!

3-2 Let us designate a lattice point in the rectangular Cartesian plane as one with integral coordinates. Consider a rectangle with sides parallel to the axes such that there are s_1 lattice points in the base and s_2 lattice points in the altitude. The vertices are lattice points.
(a) Find the number of interior lattice points, $N(I)$.
(b) Find the number of boundary lattice points, $N(B)$.
(c) Find the total number of lattice points, N.

Challenge   Suppose the word "rectangle" is changed to "square"; find $N(I)$, $N(B)$, and N.

3-3 An approximate formula for a barometric reading, p(millimeters), for altitudes h(meters) above sea level, is $p = 760 - .09h$, where $h \leq 500$. Find the change in p corresponding to a change in h from 100 to 250.

Challenge   Find the lowest and the highest pressures for which this formula is valid.

3-4 A student wishing to give 25 cents to each of several charities finds that he is 10 cents short. If, instead, he gives 20 cents to each of the charities, then he is left with 25 cents. Find the amount of money with which the student starts.

Challenge 1   How does the answer change if the original shortage is 15 cents?

Challenge 2   How does the answer change if the original shortage is 20 cents?

Challenge 3   How does the answer change if the original shortage is 25 cents?

3-5 Find two numbers x and y such that xy, $\dfrac{x}{y}$, and $x - y$ are equal.

Challenge 1   Find two numbers x and y such that $xy = \dfrac{2x}{y} = 3(x - y)$.

Challenge 2   Solve the problem for $xy = \dfrac{x}{y} = 2(x - y)$.

3-6 A merchant on his way to the market with n bags of flour passes through three tollgates. At the first gate, the toll is $\dfrac{1}{4}$ of his holdings, but 3 bags are returned. At the second gate, the toll is $\dfrac{1}{3}$ of his (new) holdings, but 2 bags are returned. At the third gate, the toll is $\dfrac{1}{2}$ of his (new) holdings, but 1 bag is returned. The merchant arrives at the market with exactly $\dfrac{n}{2}$ bags. If all transactions involve whole bags, find the value of n.

Challenge   Solve the problem if the first toll is $\dfrac{1}{6}$ of the holdings, plus $\dfrac{1}{6}$ of a bag; the second toll is $\dfrac{1}{5}$ of his (new) holdings, plus $\dfrac{1}{5}$ of a bag; and the third toll is $\dfrac{1}{4}$ of his (new) holdings, plus $\dfrac{1}{4}$ of a bag; and he arrives at the market with exactly $\dfrac{n-1}{2}$ bags.

3-7 The number N_2 is 25% more than the number N_1, the number N_3 is 20% more than N_2, and the number N_4 is x% less than N_3. For what value of x is $N_4 = N_1$?

Challenge   Solve the problem generally if N_2 is a% more than N_1, and N_3 is b% more than N_2.

3-8 Let $R = px$ represent the revenue, R (dollars), obtained from the sale of x articles, each at selling price p (dollars). Let $C = mx + b$ represent the total cost, C, in dollars, of producing and selling these x articles. How many articles must be sold to break even?

Challenge 1 What must the relation be between p and m (the unit marketing cost) to make the result meaningful? The unit marketing cost is the total marketing costs divided by the number of units marketed.

Challenge 2 How do you interpret the constant b in the given formula?

Challenge 3 Find the value of x so that the revenue exceeds the cost by $100.

Challenge 4 Find the value of x so that the revenue exceeds the cost by D dollars.

Challenge 5 Find the profit if the number of articles sold is $\dfrac{2b}{p - m}$.

3-9 In a certain examination it is noted that the average mark of those passing is 65, while the average mark of those failing is 35. If the average mark of all participants is 53, what percentage of the participants passed?

Challenge 1 Try this problem with the following changes. Replace 65 by 70, 35 by 36, but leave 53 unchanged.

Challenge 2 What is the result if the only change is 65 to 62?

3-10 Under plan I, a merchant sells n_1 articles, priced 1 for 2¢, with a profit of $\frac{1}{4}$¢ on each article, and n_2 articles, priced 2 for 3¢, with a profit of $\frac{1}{8}$¢ on each article. Under Plan II, he mixes the articles and sells them at 3 for 5¢. If $n_1 + n_2$ articles are sold under each plan, for what ratio $\dfrac{n_1}{n_2}$ is the profit the same?

Challenge Change 2¢ to p¢ and 3¢ to q¢ and solve the problem.

3-11 The sum of two numbers x and y, with $x > y$, is 36. When x is divided by 4 and y is divided by 5, the sum of the quotients is 8. Find the numbers x and y.

3-12 Find the values of x satisfying the equation $|x - a| = |x - b|$, where a, b are distinct real numbers.

Challenge 1 Find the values of x satisfying the equation $|x - 1| = |x - 2|$.

Challenge 2 Find the values of x satisfying the equation $|x + 1| = |x - 2|$.

Challenge 3 Find the values of x satisfying the equation $|2x - 1| = |x - 2|$.

Challenge 4 Find the values of x satisfying the equation $|3x - 1| = |x - 2|$.

Challenge 5 Find the values of x satisfying the equation $|2x - 1| = |2x - 2|$.

Challenge 6 Find the values of x satisfying the equation $|2x - 1| = |3x - 3|$.

Challenge 7 Now try the more difficult equation $|x - 1| + |x - 2| = |x - 3|$.

3-13 Two night watchmen, Smith and Jones, arrange for an evening together away from work. Smith is off duty every eighth evening starting today, while Jones is off duty every sixth evening starting tomorrow. In how many days from today can they get together?

Challenge Solve the problem if Smith is off duty every eighth evening starting today, while Jones is off duty alternately every sixth evening and every thirteenth evening starting tomorrow.

3-14 A man buys 3-cent stamps and 6-stamps, 120 in all. He pays for them with a $5.00 bill and receives 75 cents in change. Does he receive the correct change?

Challenge 1 Would 76 cents change be correct? Would 74 cents change be correct?

Challenge 2 If the correct change had to consist of 3 coins limited to nickels, dimes, and quarters, list the 3-coin combinations yielding an acceptable answer.

3-15 In how many ways can a quarter be changed into dimes, nickels, and cents?

Challenge Is the answer unique if it is stipulated that there are five times as many coins of one kind as of the other two kinds?

3-16 Find the number of ways in which 20 U.S. coins, consisting of quarters, dimes, and nickels, can have a value of $3.10.

4 Bases: Binary and Beyond

Prepare for a voyage to the far-out world of bases different from ten. The two main "stops" along the way are rational numbers in other bases and divisibility. You may want to read Appendix V in the back of the book before attacking the problems. It contains some unusual information on divisibility.

4-1 Can you explain mathematically the basis for the following correct method of multiplying two numbers, sometimes referred to as the Russian Peasant Method of multiplication?

Let us say that we are to find the product of 19×23. In successive rows, we halve the entries in the first column, rejecting the remainders of 1 where they occur. In the second column, we double each successive entry. This process continues until a 1 appears in column I.

I	II
19	23
9	46
4	92
2	184
1	368
	437

We then add the entries in column II, omitting those that are associated with the even entries in column I.

4-2 If $x = \{0, 1, 2, \ldots, n, \ldots\}$, find the possible terminating digits of $x^2 + x$ in base 2.

Challenge 1 If $x = \{0, 1, 2, \ldots, n, \ldots\}$, find the possible terminating digits of $x^2 + x + 1$ in base 2.

Challenge 2 If $x = \{0, 1, 2, \ldots, n, \ldots\}$, find the possible terminating digits of $x^2 + x$ in base 5.

Challenge 3 If $x = \{0, 1, 2, \ldots, n, \ldots\}$, find the possible terminating digits of $x^2 + x + 1$ in base 5.

4-3 Find the base b such that $72_b = 2(27_b)$. 72_b means 72 written in base b.

Challenge 1   Try the problem for $73_b = 2(37_b)$.

Challenge 2   Try the problem for $72_b = 3(27_b)$.

4-4 In what base b is 441_b the square of an integer?

Challenge 1   If N is the base 4 equivalent of 441 written in base 10, find the square root of N in base 4.

Challenge 2   Find the smallest base b for which 294_b is the square of an integer.

4-5 Let N be the three-digit number $a_1a_2a_3$ written in base b, $b \geq 2$, and let $S = a_1 + a_2 + a_3$. Prove that $N - S$ is divisible by $b - 1$.

Challenge 1   Prove the result for the n-digit number $a_1a_2 \ldots a_n$, written in base b.

Challenge 2   Explain the connection between this theorem and the test for divisibility by 9 in the decimal system. (See Appendix V.)

Challenge 3   Show that 73 written in base 9 is not divisible by 8, while 73 written in base 11 is divisible by 10.

4-6 Let N be the four-digit number $a_0a_1a_2a_3$ (in base 10), and let N' be the four-digit number which is any of the 24 rearrangements of the digits. Let $D = |N - N'|$. Find the largest digit that exactly divides D.

Challenge 1   Does the theorem hold for five-digit numbers? Does it hold for n-digit numbers, where n is any natural number, including single-digit numbers?

Challenge 2   Let N be the three-digit number abc with $a > c$. From N subtract the three-digit number $N' = cba$. If the digit on the left side of the difference is 4, find the complete difference.

4-7 Express in binary notation (base 2) the decimal number 6.75.

Challenge 1   Convert the decimal number $N = 19.65625$ into a binary number.

Challenge 2 Does the (base 10) non-terminating expansion 5.333 . . . terminate when converted into base 2?

4-8 Assume $r = \{6, 7, 8, 9, 10\}$ and $1 < a < r$. If there is exactly *one* integer value of a for which $\frac{1}{a}$, expressed in the base r, is a terminating r-mal, find r.

Challenge Try the problem with $\frac{5}{a}$ instead of $\frac{1}{a}$.

4-9 From the unit segment $\overline{OA}$ extending from the origin O to $A(1, 0)$, remove the middle third. Label the remaining segments $\overline{OB}$ and $\overline{CA}$, and remove the middle third from segment $\overline{OB}$. Label the first two remaining segments $\overline{OD}$ and $\overline{EB}$. Express the coordinates of D, E, and B in base 3.

Challenge Remove the middle third from segment $\overline{CA}$, and label the remaining segments $\overline{CF}$ and $\overline{FG}$. Express in base 3 the coordinates of C, F, and G.

4-10 Assume that there are n stacks of tokens with n tokens in each stack. One and only one stack consists entirely of counterfeit tokens, each token weighing 0.9 ounce. If each true token weighs 1.0 ounce, explain how to identify the counterfeit stack in one weighing, using a scale that gives a reading. You may remove tokens from any stack.

Challenge 1 Which is the counterfeit stack if the overall deficiency is $\frac{4}{5}$ ounce?

Challenge 2 What changes should be made in the analysis and solution if **(a)** each true token weighs 1.0 ounce and each counterfeit weighs 1.1 ounce? **(b)** each true token weighs 1.1 ounce and each counterfeit weighs 0.9 ounce?

Challenge 3 Solve the generalized problem of n stacks with n tokens each, if each true token weighs t ounces and each counterfeit weighs s ounces. Then apply the result to Problem 4-10 and its challenges.

5   Equations, Inequations, and Pitfalls

Equations and inequalities play a double role on the problem-solving stage. They may be the stars of the show if they are tricky to solve or especially interesting. More often, they are in the supporting cast, serving as the indispensable tool for expressing the data in a problem. Both roles are explored in this section.

5-1 Find the solution set of the equation $\dfrac{2x}{x-2} = \dfrac{4}{x-2}$.

Challenge   Find the values of x satisfying the equation $\sqrt{x-2} = -3$.

5-2 Find the pairs of numbers x, y such that $\dfrac{x-3}{2y-7} = x - 3$.

Challenge   Find the pairs (x, y) such that $\dfrac{x-3}{2y-7} = \dfrac{x-3}{2-7y}$.

5-3 Find all the real values of x such that $|\sqrt{x} - \sqrt{2}| < 1$.

Challenge   Let the set of all values of x satisfying the inequalities $|x - 8| < 6$ and $|x - 3| > 5$ be written as $a < x < b$. Find $b - a$.

5-4 Find all values of x satisfying the equation $2x = |x| + 1$.

Challenge   Compare this result with that obtained by solving the equation $2x = -|x| + 1$, and try to interpret this new equation geometrically.

5-5 Find the values of a and b so that $ax + 2 < 3x + b$ for all $x < 0$.

Challenge   Find values of a and b so that $ax + 2 \geq 3x + b$ for all $x < 0$.

5-6 Find all positive integers that leave a remainder of 1 when divided by 5, and leave a remainder of 2 when divided by 7.

Challenge 1   Change the first remainder to 2, and the second remainder to 1, and solve the problem.

Challenge 2   Solve the problem with the first remainder $1 \leq r_1 \leq 4$, and the second remainder $1 \leq r_2 \leq 6$.

5-7 On a fence are sparrows and pigeons. When five sparrows leave, there remain two pigeons for every sparrow. Then twenty-five pigeons leave, and there are now three sparrows for every pigeon. Find the original number of sparrows.

Challenge 1 Replace "five" by a and "twenty-five" by b, and find s and p (the number of sparrows and the number of pigeons, respectively).

Challenge 2 Solve the problem generally using r_1 and r_2, respectively, for the two ratios, and a and b as in Challenge 1.

5-8 A swimmer at A, on one side of a straight-banked canal 250 feet wide, swims to a point B on the other bank, directly opposite to A. His steady rate of swimming is 3 ft./sec., and the canal flow is a steady 2 ft./sec. Find the shortest time to swim from A to B.

5-9 Miss Jones buys x flowers for y dollars, where x and y are integers. As she is about to leave the clerk says, "If you buy 18 flowers more, I can let you have them all for six dollars. In this way you save 60 cents per dozen." Find a set of values for x and y satisfying these conditions.

Challenge Finding Miss Jones hesitant at the first offer, the clerk adds, "If you buy 24 flowers more, I can let you have them all for $6.75. In this way you save 75 cents per dozen." Does the same set of values for x and y satisfy these new equations?

5-10 Find the set of real values of x satisfying the equation

$$\frac{x+5}{x+4} - \frac{x+6}{x+5} = \frac{x+7}{x+6} - \frac{x+8}{x+7}.$$

Challenge 1 After solving the problem can you find, by inspection, the answer to

$$\frac{x+6}{x+5} - \frac{x+7}{x+6} = \frac{x+8}{x+7} - \frac{x+9}{x+8}?$$

Challenge 2 Solve the more general problem

$$\frac{x+a+1}{x+a} - \frac{x+a+2}{x+a+1} = \frac{x+a+3}{x+a+2} - \frac{x+a+4}{x+a+3}.$$

5-11 The contents of a purse are not revealed to us, but we are told that there are exactly 6 pennies and at least one nickel and one dime. We are further told that if the number of dimes were changed to the number of nickels, the number of nickels were changed to the number of pennies, and the number of pennies were changed to the number of dimes, the sum would remain unchanged. Find the least possible and the largest possible number of coins the purse contains.

Challenge 1   How does the situation change if the number of nickels is 6, and the number of dimes and the number of pennies are unspecified, except that there must be at least one of each?

Challenge 2   What solution is obtained if the number of dimes is 6, but the nickels and pennies are unspecified?

Challenge 3   Explain why, in the original problem, the least number of coins yields the greatest value, whereas in Challenges 1 and 2 the least number of coins yields the smallest value.

Challenge 4   Investigate the problem if there are exactly 6 pennies, and at least one nickel, one dime, and one quarter.

5-12 A shopper budgets twenty cents for twenty hardware items. Item A is priced at 4 cents each, item B, at 4 for 1 cent, and item C, at 2 for 1 cent. Find all the possible combinations of 20 items made up of items A, B, and C that are purchasable.

5-13 Partition 75 into four positive integers a, b, c, d such that the results are the same when 4 is added to a, subtracted from b, multiplied by c, and divided into d. To partition a positive integer is to represent the integer as a sum of positive integers.

Challenge 1   Partition 48 into four parts a, b, c, d such that the results are the same when 3 is added to a, subtracted from b, multiplied by c, and divided by d.

Challenge 2   Partition 100 into five parts a, b, c, d, e so that the results are the same when 2 is added to a, 2 is subtracted from b, 2 is multiplied by c, 2 is divided by d, and the positive square root is taken of e.

5-14 Two trains, each traveling uniformly at 50 m.p.h., start toward each other, at the same time, from stations A and B, 10 miles apart. Simultaneously, a bee starts from station A, flying parallel to the track at the uniform speed of 70 m.p.h., toward the train from station B. Upon reaching the train it comes to rest, and allows itself to be transported back to the point where the trains pass each other. Find the total distance traveled by the bee.

5-15 One hour out of the station, the locomotive of a freight train develops trouble that slows its speed to $\frac{3}{5}$ of its average speed up to the time of the failure. Continuing at this reduced speed it reaches its destination two hours late. Had the trouble occurred 50 miles beyond, the delay would have been reduced by 40 minutes. Find the distance from the station to the destination.

5-16 Two trains, one 350 feet long, the other 450 feet long, on parallel tracks, can pass each other completely in 8 seconds when moving in opposite directions. When moving in the same direction, the faster train completely passes the slower one in 16 seconds. Find the speed of the slower train.

Challenge 1 Show that, if the respective times are t_1 and t_2 with $t_2 > t_1$, the results are

$$f = \frac{800(t_2 + t_1)}{2t_2 t_1} \text{ and } s = \frac{800(t_2 - t_1)}{2t_2 t_1}.$$

Challenge 2 Show that, if the respective times are t_1 and t_2 with $t_2 > t_1$, and the respective lengths are L_1 and L_2, the results are

$$f = \frac{(L_1 + L_2)(t_2 + t_1)}{2t_2 t_1} \text{ and } s = \frac{(L_1 + L_2)(t_2 - t_1)}{2t_2 t_1}.$$

Use this formula to solve the original problem.

5-17 The equation $5(x - 2) = \frac{27}{7}(x + 2)$ is written throughout in base 9. Solve for x, expressing its values in base 10.

5-18 Find the two prime factors of 25,199 if one factor is about twice the other.

Challenge   Find the three prime factors of 27,931 if the three factors are approximately in the ratio 1:2:3.

5-19   When asked the time of the day, a problem-posing professor answered, "If you add one-eighth of the time from noon until now to one-quarter the time from now until noon tomorrow, you get the time exactly." What time was it?

Challenge 1   On another occasion the professor said, "If from the present time, you subtract one-sixth of the time from now until noon tomorrow, you get exactly one-third of the time from noon until now." Find the present time.

Challenge 2   If, as the result of daylight-saving time confusion, the professor's watch is one hour fast, find the change needed in the original statement "one-eighth of the time from noon until now" to yield the answer 5:20 P.M. true time.

Challenge 3   One day the professor forgot his watch. A colleague, of whom he asked the time, in an attempt to cure the professor of his mannerism, replied, "If you subtract two-thirds the time from now until noon tomorrow from twice the time from noon to now, you get the time short by ten minutes."

Do you agree with the professor that it was 9:30 P.M.?

5-20   Solve $\dfrac{1}{x} + \dfrac{1}{y} = \dfrac{1}{z}$ for integer values of x, y, and z.

5-21   Prove that, for the same set of integral values of x and y, both $3x + y$ and $5x + 6y$ are divisible by 13.

6 Correspondence: Functionally Speaking

Finding a particular value of a function or its range of values is ordinarily a routine task. But not for the functions considered in this section. In addition to functions defined algebraically, there are functions defined in terms of geometric ideas such as paths along a coordinate grid and partitions of the plane by families of lines.

6-1 Define the symbol $f(a)$ to mean the value of a function f of a variable n when $n = a$. If $f(1) = 1$ and $f(n) = n + f(n - 1)$ for all natural numbers $n \geq 2$, find the value of $f(6)$.

Challenge 1 Find $f(8)$ by using the method of "telescopic" addition and, if possible, by using a short cut (see Challenge 2).

Challenge 2 Show that $f(n) = \frac{1}{2}n(n + 1)$.

Challenge 3 Find the value of $f(10)$; of $f(100)$.

Challenge 4 Find the value of n such that $f(n) = 3f(5)$.

Challenge 5 Find the value of n such that $f(4n) = 12f(n)$.

6-2 Each of the following (partial) tables has a function rule associating a value of n with its corresponding value $f(n)$. If $f(n) = An + B$, determine for each case the numerical values of A and B.

(a)

n	$f(n)$
$\frac{1}{2}$	$\frac{1}{2}$
$\frac{1}{3}$	$\frac{2}{3}$
$\frac{3}{5}$	$\frac{2}{5}$

(b)

n	$f(n)$
$\frac{1}{2}$	$\frac{3}{2}$
$\frac{1}{3}$	$\frac{4}{3}$
$\frac{2}{5}$	$\frac{7}{5}$

(c)

n	$f(n)$
$\frac{1}{2}$	0
$\frac{1}{3}$	$\frac{1}{3}$
$\frac{1}{4}$	$\frac{1}{2}$

(d)

n	$f(n)$
$2\frac{3}{4}$	$3\frac{1}{2}$
$2\frac{1}{4}$	3
1	$1\frac{3}{4}$

(e)

n	$f(n)$
$2\frac{3}{4}$	2
$2\frac{1}{4}$	$1\frac{1}{2}$
1	$\frac{1}{4}$

Challenge Now try a slightly harder table.

n	$1\frac{1}{2}$	1	$\frac{1}{2}$
$f(n)$	0	-1	-2

6-3 In a given right triangle, the perimeter is 30 and the sum of the squares of the sides is 338. Find the lengths of the three sides.

Challenge Redo the problem using an area of 30 in place of the perimeter of 30.

6-4 A rectangular board is to be constructed to the following specifications:

(a) the perimeter is equal to or greater than 12 inches, but less than 20 inches

(b) the ratio of adjacent sides is greater than 1 but less than 2. Find all sets of integral dimensions satisfying these specifications.

6-5 Find the range of values of $F = \dfrac{x^2}{1 + x^4}$ for real values of x.

Challenge Find the largest and the least values of $f = \dfrac{2x + 3}{x + 2}$ for $x \geq 0$.

6-6 Determine the largest possible value of the function $x + 4y$ under the four conditions: (1) $5x + 6y \leq 30$ (2) $3x + 2y \leq 12$ (3) $x \geq 0$ (4) $y \geq 0$.

6-7 Let us define the distance from the origin O to point A as the length of the *path along the coordinate lines*, as shown in Fig. 6-7, so that the distance from O to A is 3.

Starting at O, how many points can you reach if the distance, as here defined, is n, where n is a positive integer?

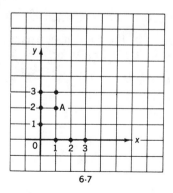

6-7

6-8 Given n straight lines in a plane such that each line is infinite in extent in both directions, no two lines are parallel (fail to meet), and no three lines are concurrent (meet in one point), into how many regions do the n lines separate the plane?

Challenge 1 Let there be $n = r + k$ lines in the plane (infinite in both directions) such that no three of the n lines are concurrent, but k lines are parallel (but no others). Find the number of partitions of the plane.

Challenge 2 Let there be n straight lines in the plane (infinite in both directions) such that three (and only three) are concurrent and such that no two are parallel. Find the number of plane separations.

Challenge 3 A set of k_1 parallel lines in the plane is intersected by another set of k_2 parallel lines, all infinite in extent. Find the number of plane separations.

Challenge 4 In Challenge 3, introduce an additional line not parallel to any of the given lines, and not passing through any of the $k_1 k_2$ points of intersection. How many additional regions are created by this plane? What is the total number of plane separations?

Challenge 5 In a given plane, let there be n straight lines, infinite in extent, four of which, and only four, are concurrent and no two of which are parallel. Find the number of planar regions.

6-9 Define $\frac{N}{D}$ as a proper fraction when $\frac{N}{D} < 1$ with N, D natural numbers. Let $f(D)$ be the number of irreducible proper fractions with denominator D. Find $f(D)$ for $D = 51$.

Challenge Find $f(D)$ for $D = 52$.

7   Equations and Inequations: Traveling in Groups

In one way or another, the problems in this section concern systems of equations or inequalities (or both!). Two unusual topics included are the notion of an approximate solution for an inconsistent system of equations and linear programming.

7-1 Let the lines $15x + 20y = -2$ and $x - y = -2$ intersect in point P. Find all values of k which ensure that the line $2x + 3y = k^2$ goes through point P.

7-2 Let (x, y) be the coordinates of point P in the xy-plane, and let (X, Y) be the coordinates of point Q (the image of point P) in the XY-plane. If $X = x + y$ and $Y = x - y$, find the simplest equation for the set of points in the XY-plane which is the image of the set of points $x^2 + y^2 = 1$ in the xy-plane.

7-3 The numerator and the denominator of a fraction are integers differing by 16. Find the fraction if its value is more than $\frac{5}{9}$ but less than $\frac{4}{7}$.

7-4 If $x + y + 2z = k$, $x + 2y + z = k$, and $2x + y + z = k$, $k \neq 0$, express $x^2 + y^2 + z^2$ in terms of k.

Challenge 1   Solve the problem if $x + 2y + 3z = k$, $3x + y + 2z = k$, and $2x + 3y + z = k$, $k \neq 0$.

Challenge 2   Show that both the problem and Challenge 1 can be solved by inspection.

7-5 Why are there no integer solutions of $x^2 - 5y = 27$?

7-6 Civic Town has 500 voters, all of whom vote on two issues in a referendum. The first issue shows 375 in favor, and the second issue shows 275 in favor. If there are exactly 40 votes against both issues, find the number of votes in favor of both issues.

7-7 How do you find the true weight of an article on a balance scale in which the two arms (distances from the pans to the point of support) are unequal?

Challenge Suppose it is known that the arms of a balance scale are unequal; how do you determine the ratio r of the arm lengths?

7-8 Solve in base 7 the pair of equations $2x - 4y = 33$ and $3x + y = 31$, where x, y, and the coefficients are in base 7.

7-9 Given the pair of equations $2x - 3y = 13$ and $3x + 2y = b$, where b is an integer and $1 \leq b \leq 100$, let $n^2 = x + y$, where x and y are integer solutions of the given equations associated in proper pairs. Find the positive values of n for which these conditions are met.

7-10 Find the set of integer pairs satisfying the system

$$3x + 4y = 32$$
$$y > x$$
$$y < \frac{3}{2}x.$$

7-11 Compare the solution of system I,

$$x + y + z = 1.5$$
$$2x - y + z = 0.8$$
$$x - 2y - z = -0.2,$$

with that of system II,

$$x + y + z = 1.5$$
$$2x - y + z = 0.9$$
$$x - 2y - z = -0.2.$$

7-12 The following information was obtained by measurement in a series of experiments:

$$x + y = 1.9$$
$$2x - y = 1.4$$
$$x - 2y = -0.6$$
$$x - y = 0.3.$$

Find an approximate solution to this system of equations.

7-13 Find the maximum and the minimum values of the function $3x - y + 5$, subject to the restrictions $y \geq 1$, $x \leq y$, and $2x - 3y + 6 \geq 0$.

7-14 A buyer wishes to order 100 articles of three types of merchandise identified as A, B, and C, each costing \$5, \$6, and \$7, respectively. From past experience, he knows that the number of each article bought should not be less than 10 nor more than 60, and that the number of B articles should not exceed the number of A articles by more than 30. If the selling prices for the articles are \$10 for A, \$15 for B, and \$20 for C, and all the articles are sold, find the number of each article to be bought so that there is a maximum profit.

---

8   Miscellaneous: Curiosity Cases

The mixed bag of problems in this final section unites several themes from earlier sections. Many problems deal with topics considered before. More important, the solutions involve techniques that have been illustrated in previous sections.

8-1 Find all values of x satisfying the equation $\dfrac{x - \sqrt{x+1}}{x + \sqrt{x+1}} = \dfrac{11}{5}$.

8-2 Find all real values of x satisfying the equations:

(a) $x^2|x| = 8$

(b) $x|x^2| = 8$, where the symbol $|x|$ means $+x$ when $x \geq 0$, and $-x$ when $x < 0$.

Challenge   Replace $+8$ by -8 in each equation and find the real values of x satisfying the new equations.

8-3 Let $P(x, y)$ be a point on the graph of $y = x + 5$. Connect P with $Q(7, 0)$. Let a perpendicular from P to the x-axis intersect it in R. Restricting the abscissa of P to values between 0 and 7, both included, find:

(a) the maximum area of right triangle PRQ

(b) two positions of P yielding equal areas.

8-4 Find the smallest value of x satisfying the conditions: $x^3 + 2x^2 = a$, where x is an odd integer, and a is the square of an integer.

Challenge 1 Change "odd integer" to "even integer greater than 2," and solve the problem.

Challenge 2 Change $x^3 + 2x^2$ to $x^3 - 2x^2$, and then solve the problem.

Challenge 3 In Challenge 1 change $x^3 + 2x^2$ to $x^3 - 2x^2$ and solve the problem.

8-5 If $\frac{3x-5}{x^2-1} = \frac{A}{x-1} + \frac{B}{x+1}$ is true for all permissible values of x, find the numerical value of $A + B$.

Challenge Solve the problem with B as the first numerator on the right-hand side of the equation, and A as the second numerator.

8-6 For what integral values of x and p is $(x^2 - x + 3)(2p + 1)$ an odd number?

Challenge Solve the problem using $(x^2 + x + 3)(2p - 1)$.

8-7 Express the simplest relation between a, b, and c, not all equal, if $a^2 - bc = b^2 - ca = c^2 - ab$.

Challenge Solve the problem for $a^2 + bc = b^2 + ca = c^2 + ab$.

8-8 Find the two linear factors with integral coefficients of $P(x, y) = x^2 - 2y^2 - xy - x - y$, or show that there are no such factors.

Challenge 1 Change $P(x, y)$ to $x^2 - 2y^2 - xy - 2x - 5y - 3$, and find the linear factors with integral coefficients.

Challenge 2 Change $P(x, y)$ to $x^2 - 2y^2 - xy - 2x - 5y + 3$, and show that linear factors with integral coefficients do not exist.

8-9 Find the sum of the digits of $(100{,}000 + 10{,}000 + 1000 + 100 + 10 + 1)^2$.

8-10 How do you invert a fraction, using the operation of addition?

8-11 How do you generate the squares of integers from pairs of consecutive integers?

8-12 Is there an integer N such that $N^3 = 9k + 2$, where k is an integer?

Challenge   Is there an integer N such that $N^3 = 9k + 8$?

8-13 Let $S_n = 1^n + 2^n + 3^n + 4^n$, and let $S_1 = 1 + 2 + 3 + 4 = 10$. Show that S_n is a multiple of S_1 for all natural numbers n, except $n = 4k$, where $k = 0, 1, 2, \ldots$.

8-14 A positive integer N is squared to yield N_1, and N_1 is squared to yield N_2. When N_2 is multiplied by N the result is a seven-digit number ending in 7. Find N.

8-15 Let $f = mx + ny$, where m, n are fixed positive integers, and x, y are positive numbers such that xy is a fixed constant. Find the minimum value of f.

SECTION II
Second Year Algebra

---

9 Diophantine Equations: The Whole Answer

An equation with two or more variables whose values are restricted to integers is known as a Diophantine equation after Diophantus of Alexandria, who studied them about 1800 years ago. They may arise in describing situations involving objects that occur only in integral quantities. Some of the problems here, for example, concern people, coins, or pieces of merchandise. A solution of a Diophantine equation is an ordered pair (or triple, or quadruple, etc.) of integers. When solutions exist, there are generally an infinite number of them. If further restrictions are imposed on the values of the variables, such as that they must be positive or less than a certain integer, there may be a finite number of solutions or even just one.

9-1 A shopkeeper orders 19 large and 3 small packets of marbles, all alike. When they arrive at the shop, he finds the packets broken open with all the marbles loose in the container. Can you help the shopkeeper make new packets with the proper number of marbles in each, if the total number of marbles is 224?

Challenge Redo the problem with 19 small packets and 3 large packets.

9-2 Find the integral solutions of $6x + 15y = 23$.

Challenge 1 Solve in positive integers $13x + 21y = 261$.

Challenge 2 Show that there are no positive integral solutions of $17x + 15y = 5$ but that $17x - 15y = 5$ has infinitely many positive integral solutions.

9-3 A picnic group transported in n buses (where $n > 1$ and not prime) to a railroad station, together with 7 persons already waiting at the station, distribute themselves equally in 14 railroad cars. Each bus, nearly filled to its capacity of 52 persons, carried the same number of persons. Assuming that the number of picnickers is the smallest possible for the given conditions, find the number of persons in each railroad car.

Challenge   Solve the problem with the following changes:

(a) 11 persons are waiting at the station instead of 7.

(b) There are 22 railroad cars and each of 21 cars has the same number of persons, but in the 22nd car there are 10 vacant seats.

9-4 Find the number of ways that change can be made of $1.00 with 50 coins (U.S.).

Challenge   Solve the problem restricting the change to dimes, nickels, and cents.

9-5 Let x be a member of the set $\{1, 2, 3, 4, 5, 6, 7\}$, y, a member of the set $\{8, 9, 10, 11, 12, 13, 14\}$, and z, a member of the set $\{15, 16, 17, 18, 19, 20, 21\}$. If a solution of $x + y + z = 33$ is defined as a triplet of integers, one each for x, y, and z taken from their respective sets, find the number of solutions.

Challenge   Solve the problem for $x + y + z = 31$.

9-6 R_1, R_2, and R_3 are three rectangles of equal area. The length of R_1 is 12 inches more than its width, the length of R_2 is 32 inches more than its width, and the length of R_3 is 44 inches more than its width. If all dimensions are integers, find them.

9-7 Given $x^2 = y + a$ and $y^2 = x + a$ where a is a positive integer, find expressions for a that yield integer solutions for x and y.

9-8 A merchant has six barrels with capacities of 15, 16, 18, 19, 20, and 31 gallons. One barrel contains liquid B, which he keeps for himself, the other five contain liquid A, which he sells to two men so that the quantities sold are in the ratio $1:2$. If none of the barrels is opened, find the capacity of the barrel containing liquid B.

9-9 Find the number of ordered pairs of integer solutions (x, y) of the equation $\dfrac{1}{x} + \dfrac{1}{y} = \dfrac{1}{p}$, p a positive integer.

9-10 Express in terms of A the number of solutions in positive integers of $x + y + z = A$ where A is a positive integer greater than 3.

9-11 Solve in integers $ax + by = c$ where a, b, and c are integers, $a < b$, and $1 \leq c \leq b$, with a and b relatively prime.

10 Functions: A Correspondence Course

A number of "advanced" ideas concerning functions will be encountered in this section. Among them are recursive definitions, limits of functions, and composite functions. A few strange exponential functions make an appearance, as do some unusual symmetric functions. The notions of variation and continued fractions complete the cast of characters.

10-1 Let f be defined as $f(3n) = n + f(3n - 3)$ when n is a positive integer greater than 1, and $f(3n) = 1$ when $n = 1$. Find the value of $f(12)$.

Challenge Define f to be such that $f(3n) = n^2 + f(3n - 3)$. Find $f(15)$.

10-2 If f is such that $f(x) = 1 - f(x - 1)$, express $f(x + 1)$ in terms of $f(x - 1)$.

Challenge 1 Does the constant function $f = \dfrac{1}{2}$ satisfy these conditions?

Challenge 2 Express $f(x + 2)$ in terms of $f(x - 1)$.

Challenge 3 Express $f(x + n)$ in terms of $f(x - 1)$.

10-3 Let $f = ax + b$, $g = cx + d$, x a real number, a, b, c, d real constants. **(a)** Find relations between the coefficients so that $f(g)$ is identically equal to x; that is, $f(g) \equiv x$, and **(b)** show that, when $f(g) \equiv x$, $f(g)$ implies $g(f)$.

Challenge 1 Prove that the results are the same for $f = ax - b$ and $g = cx - d$.

Challenge 2 Solve the problem for $f = ax + b$ and $g = cx - d$.

10-4 If $f(x) = -x^n(x - 1)^n$, find $f(x^2) + f(x)f(x + 1)$.

Challenge Find $f(x^4) + f(x^2)f(x^2 + 1)$ by inspection.

10-5 The density d of a fly population varies directly as the population N, and inversely as the volume V of usable free space. It is also determined experimentally that the density for a maximum population varies directly as V. Express N (maximum) in terms of V.

Challenge Solve the problem if the density for a maximum population varies directly as $\sqrt{V}$.

10-6 Given the four elementary symmetric functions $f_1 = x_1 + x_2 + x_3 + x_4$, $f_2 = x_1x_2 + x_2x_3 + x_3x_4 + x_4x_1$, $f_3 = x_1x_2x_3 + x_2x_3x_4 + x_3x_4x_1 + x_4x_1x_2$, $f_4 = x_1x_2x_3x_4$, express $S = \dfrac{1}{x_1} + \dfrac{1}{x_2} + \dfrac{1}{x_3} + \dfrac{1}{x_4}$ in terms of f_1, f_2, f_3, f_4.

Challenge Let $g_1 = x_1 + x_2 + x_3$, $g_2 = x_1x_2 + x_2x_3 + x_3x_1$, $g_3 = x_1x_2x_3$. Express $T = \dfrac{1}{x_1{}^2} + \dfrac{1}{x_2{}^2} + \dfrac{1}{x_3{}^2}$ in terms of g_1, g_2, g_3.

10-7 Let $f(n) = n(n + 1)$ where n is a natural number. Find the values of m and n such that $4f(n) = f(m)$ where m is a natural number.

Challenge 1 Try the problem with $2f(n) = f(m)$.

Challenge 2 Try the problem with $5f(n) = f(m)$.

10-8 Find the positive real values of x such that $x^{(x^x)} = (x^x)^x$.

Challenge For which positive values of x is $x^{(x^x)} > (x^x)^x$, and for which positive values of x is $x^{(x^x)} < (x^x)^x$?

10-9 If $x = 3 + \cfrac{1}{3 + \cfrac{1}{x}}$ and $y = 3 + \cfrac{1}{3 + \cfrac{1}{3 + \cfrac{1}{y}}}$, find the value of $|x - y|$.

Challenge Express $\dfrac{3 + \sqrt{13}}{2}$ as an (infinite) continued fraction. See problem 10-10.

10-10 Assuming that the infinite continued fraction $\cfrac{2}{2 + \cfrac{2}{2 + \cfrac{2}{2 + \cdots}}}$

represents a finite value x, find x. (Technically, we say the infinite continued fraction converges to the value x.)

Challenge 1 Assuming convergence, find
$$y = -2 + \cfrac{2}{-2 + \cfrac{2}{-2 + \cfrac{2}{-2 + \cdots}}}.$$

Challenge 2 Assuming convergence, find $y = 1 + \cfrac{1}{1 + \cfrac{1}{1 + \cfrac{1}{1 + \cdots}}}.$

10-11 Find $\lim\limits_{h \to 0} F$; that is, the limiting value of F as h becomes arbitrarily close to zero where $F = \dfrac{\sqrt{3 + h} - \sqrt{3}}{h}$, $h \neq 0$.

Challenge 1 Find $\lim\limits_{h \to 0} G$ where $G = \dfrac{\sqrt{2 + h} - \sqrt{2}}{h}$, $h \neq 0$.

Challenge 2 Find $\lim\limits_{h \to 0} H$ where $H = \dfrac{\sqrt{4 + h} - \sqrt{4}}{h}$, $h \neq 0$.

Challenge 3 Find $\lim\limits_{h \to 0} X$ where $X = \dfrac{\sqrt{x + h} - \sqrt{x}}{h}$, $h \neq 0$.

10-12 Find the limiting value of $F = \dfrac{x^a - 1}{x - 1}$, $x \neq 1$, where a is a positive integer, as x assumes values arbitrarily close to 1; that is, find $\lim\limits_{x \to 1} F$.

10-13 If n is a real number, find $\lim\limits_{x \to 2} \dfrac{x^n - 2^n}{x - 2}$ in terms of n, that is, the limiting value of $\dfrac{x^n - 2^n}{x - 2}$, as x approaches arbitrarily close to 2.

Challenge   On the basis of the results obtained here and in Problem 10-12 find, by inspection, $\lim\limits_{x\to 3}\dfrac{x^n - 3^n}{x - 3}$ and $\lim\limits_{x\to a}\dfrac{x^n - a^n}{x - a}$, where a is a positive integer.

10-14 A function f is defined as

$$f = \begin{cases} 1, \text{ when } x = 1, \\ 2x - 1 + f(x - 1), \text{ when } x \geq 2, x \text{ an integer.} \end{cases}$$

Express f as the simplest possible polynomial.

Challenge 1   Solve the problem for

$$f = \begin{cases} 1, \text{ when } x = 0, \\ 2x + 1 + f(x - 1), \text{ when } x \geq 1, x \text{ an integer.} \end{cases}$$

Challenge 2   Solve the problem for

$$f = \begin{cases} 1, \text{ when } x = 0, \\ 2x - 1 + f(x - 1), \text{ when } x \geq 1, x \text{ an integer.} \end{cases}$$

11   Inequalities, More or Less

As a major means for expressing mathematical comparisons, inequalities have become a key tool in modern mathematics. Several problems in this section involve comparing the size of two numbers, while others call for solving a tricky inequality. Various applications of inequalities are made in the remaining problems.

11-1 Let $P = \left(\dfrac{1}{a} - 1\right)\left(\dfrac{1}{b} - 1\right)\left(\dfrac{1}{c} - 1\right)$, where a, b, c are positive numbers such that $a + b + c = 1$. Find the largest integer N such that $P \geq N$.

11-2 Find the pair of least positive integers x and y such that $11x - 13y = 1$ and $x + y > 50$.

Challenge 1   Replace $x + y > 50$ by $x + y \geq 59$ and solve the problem.

Challenge 2 Replace $x + y > 50$ by $x + y > 59$ and solve the problem.

11-3 Is the following set of inequalities consistent? (Consider three inequalities at a time.)

$$x + y \leq 3, \quad -x - y \geq 0, \quad x \geq -1, \quad -y \leq 2$$

11-4 Find the set of values for x such that $x^3 + 1 > x^2 + x$.

Challenge Find the set of values of x such that $x^3 - 1 > x^2 - x$.

11-5 Consider a triangle whose sides a, b, c have integral lengths such that $c < b$ and $b \leq a$. If $a + b + c = 13$ (inches), find all the possible distinct combinations of a, b, and c.

Challenge Change the perimeter from 13 to 15 (inches) and solve the problem.

11-6 A teen-age boy is now n times as old as his sister, where $n > 3\frac{1}{2}$. In 3 years he will be $n - 1$ times as old as she will be then. If the sister's age, in years, is integral, find the present age of the boy.

Challenge Solve the problem when $3 < n < 4$.

11-7 Express the maximum value of A in terms of n so that the following inequality holds for any positive integer n.

$$x^n + x^{n-2} + x^{n-4} + \cdots + \frac{1}{x^{n-4}} + \frac{1}{x^{n-2}} + \frac{1}{x^n} \geq A$$

11-8 Find the set $R_1 = \{x \mid x^2 + (x^2 - 1)^2 \geq |2x(x^2 - 1)|\}$, and the set $R_2 = \{x \mid x^2 + (x^2 - 1)^2 < |2x(x^2 - 1)|\}$.

Challenge Replace $x^2 + (x^2 - 1)^2$ by $x^2 + (x - 1)^2$ and $2x(x^2 - 1)$ by $2x(x - 1)$ and solve the problem.

11-9 Show that $F = \dfrac{1}{2} \cdot \dfrac{3}{4} \cdot \dfrac{5}{6} \cdots \dfrac{99}{100} < \dfrac{1}{\sqrt{101}}$.

Challenge Show that $P = \dfrac{2}{3} \cdot \dfrac{4}{5} \cdot \dfrac{6}{7} \cdots \dfrac{100}{101} > \dfrac{\sqrt{101}}{101}$.

11-10 Which is larger $\sqrt[9]{9!}$ or $\sqrt[10]{10!}$? Be careful!

11-11 If x is positive, how large must x be so that $\sqrt{x^2 + x} - x$ shall differ from $\frac{1}{2}$ by less than 0.02?

Challenge Solve the problem with 0.02 replaced by 0.01.

11-12 Find a rational approximation $\frac{m}{n}$ to $\sqrt{2}$ such that $-\frac{1}{8n} < \sqrt{2} - \frac{m}{n} < \frac{1}{8n}$ where $n \leq 8$.

11-13 Find the least value of $(a_1 + a_2 + a_3 + a_4)\left(\frac{1}{a_1} + \frac{1}{a_2} + \frac{1}{a_3} + \frac{1}{a_4}\right)$, where each a_i, $i = 1, 2, 3, 4$, is positive.

Challenge 1 When are the factors equal?

Challenge 2 Verify the theorem for $a_1 = a_2 = \cdots = a_{n-1} = 1$ and $a_n = 2$.

12 Number Theory: Divide and Conquer

Though many exotic branches of mathematics have flowered since the turn of the century, some of the deepest mathematical research in progress today concerns that oldtimer, number theory. The problems in this section convey the flavor of this persistently vital subject. They deal with such topics as divisibility, factorization, number bases, and congruence.

12-1 Let $N_1 = .888 \ldots$, written in base 9, and let $N_2 = .888 \ldots$, written in base 10. Find the value of $N_1 - N_2$ in base 9.

Challenge 1 Express $N_1 - N_2$ in base 10.

Challenge 2 Express $N_1 - N_2$ in base 12.

12-2 Solve $x^2 - 2x + 2 \equiv 0 \pmod 5$.

Challenge 1 Solve $x^2 - 2x + 2 \equiv 0 \pmod{10}$.

Challenge 2 Solve $x^2 - 2x + 2 \equiv 0 \pmod{17}$.

12-3 Find the positive digit divisors, other than 1, of $N = 664,512$ written in base 9.

Challenge Is $N = 664,426$ written in base 8 divisible by 7?

12-4 Find all the positive integral values of n for which $n^4 + 4$ is a prime number.

Challenge Solve the problem for $n^4 + n^2 + 1$.

12-5 Let $B_a = x^a - 1$ and let $B_b = x^b - 1$ with a, b positive integers. If $B_y = x^y - 1$ is the binomial of highest degree dividing each of B_a and B_b, how is y related to a and b?

Challenge Does the result hold for $x^a + 1$ and $x^b + 1$?

12-6 If $f(x) = x^4 + 3x^3 + 9x^2 + 12x + 20$, and $g(x) = x^4 + 3x^3 + 4x^2 - 3x - 5$, find the functions $a(x)$, $b(x)$ of smallest degree such that $a(x)f(x) + b(x)g(x) = 0$.

12-7 Find the smallest positive integral value of k such that $kt + 1$ is a triangular number when t is a triangular number. (See Appendix VII.)

12-8 Express the decimal .3 in base 7.

Challenge 1 Express the decimal .4 in base 7.

Challenge 2 Express the decimal .5 in base 7.

12-9 The following excerpt comes from Lewis Carroll's *Alice's Adventures in Wonderland*.

"Let me see: four times five is twelve, and four times six is thirteen, and four times seven is—oh dear! I shall never get to twenty at that rate!"

Do you agree or disagree with the author?

12-10 Show that, if $a^2 + b^2 = c^2$, a, b, c integers, then $P = abc$ is divisible by $60 = 3 \cdot 4 \cdot 5$.

12-11 Find the integer values of x between -10 and $+15$ such that $P = 3x^3 + 7x^2$ is the square of an integer.

Challenge What is the least value of $x > 15$ satisfying the given conditions?

12-12 Find the geometric mean of the positive divisors of the natural number n. (See Appendix IV.)

12-13 Show that if $P = 1 \cdot 2 \cdot 3 \cdot \cdots n$ and $S = 1 + 2 + 3 + \cdots + n$, n a natural number, then S exactly divides P if n is odd.

12-14 By shifting the initial digit 6 of the positive integer N to the end, we obtain a number equal to $\frac{1}{4} N$. Find the smallest possible value of N that satisfies the conditions.

Challenge Solve the problem with initial digit 8.

12-15 Find the two-digit number N (base 10) such that when it is divided by 4 the remainder is zero, and such that all of its positive integral powers end in the same two digits as the number.

12-16 Find a base b such that the number 321_b (written in base b) is the square of an integer written in base 10.

Challenge Find a base b such that 123_b is the cube of an integer written in base 10.

12-17 If $\dfrac{(a-b)(c-d)}{(b-c)(d-a)} = -\dfrac{5}{3}$, find $\dfrac{(a-c)(b-d)}{(a-b)(c-d)}$.

Challenge If $\dfrac{(a+b)(c+d)}{(b+c)(d+a)} = -\dfrac{5}{3}$, find $\dfrac{(a-c)(b-d)}{(a+b)(c+d)}$.

12-18 Solve $x(x+1)(x+2)(x+3) + 1 = y^2$ for integer values of x and y.

Challenge Solve $(x+1)(x+2)(x+3)(x+4) + 1 = y^2$ for integer values of x and y.

12-19 Factor $x^4 - 6x^3 + 9x^2 + 100$ into quadratic factors with integral coefficients.

Challenge Find the quadratic factors with integral coefficients of $x^4 - 10x^3 + 10x^2 - 41x - 20$.

12-20 Express $(a^2 + b^2)(c^2 + d^2)$ as the sum of the squares of two binomials in four ways.

12-21 Observe that 1234 is not divisible by 11, but a rearrangement (permutation) of the digits such as 1243 is divisible by 11. Find the total number of permutations that are divisible by 11.

Challenge   Find the total number of permutations of $N = 12345$ that are divisible by 11.

12-22 Find all integers N with initial (leftmost) digit 6 with the property that, when the initial digit is deleted, the resulting number is $\frac{1}{16}$ of the original number N.

Challenge   Solve the problem with initial digit 9.

12-23 Find the largest positive integer that exactly divides $N = 11^{k+2} + 12^{2k+1}$, where $k = 0, 1, 2, \ldots$.

Challenge 1   Find the largest positive integer exactly dividing $N = 7^{k+2} + 8^{2k+1}$, where $k = 0, 1, 2, \ldots$.

Challenge 2   Show in general terms that $N = A^{k+2} + (A + 1)^{2k+1}$, where $k = 0, 1, 2, \ldots$, is divisible by $(A + 1)^2 - A$.

12-24 For which positive integral values of x, if any, is the equation $x^6 = 9k + 1$, where $k = 0, 1, 2, \ldots$, not satisfied?

12-25 If n, A, B, and C are positive integers, and $A^n - B^n - C^n$ is divisible by BC, express A in terms of B and C (free of n).

12-26 Prove that if $ad = bc$, then $P = ax^3 + bx^2 + cx + d, a \neq 0$, is divisible by $x^2 + h^2$, where $h^2 = \frac{c}{a} = \frac{d}{b}$.

Challenge   Try to prove the converse of this theorem.

12-27 Let R be the sum of the reciprocals of all positive factors, used once, of N, including 1 and N, where $N = 2^{p-1}(2^p - 1)$, with $2^p - 1$ a prime number. Find the value of R.

12-28 Note that $180 = 3^2 \cdot 20 = 3^2 \cdot 2^2 \cdot 5$ can be written as the sum of two squares of integers, namely, $36 + 144 = 6^2 + 12^2$, but that $54 = 3^2 \cdot 6 = 3^2 \cdot 2 \cdot 3$ cannot be so expressed. If a, b are integers, find the nature of the factor b such that $a^2 \cdot b$ is the sum of two squares of integers.

12-29 Show that $b - 1$ divides $b^{b-2} + b^{b-3} + \cdots + b + 1$, and thus show that $b^2 - 2b + 1$ divides $b^{b-1} - 1$.

---

13 Maxima and Minima: Ups and Downs

Since ancient times, mathematicians have been continually fascinated by the idea of maximizing and minimizing mathematical objects or quantities. There seems to be something especially irresistible about the challenge of finding the largest rectangle that can be inscribed in a certain triangle or the smallest value of a certain function. You will find a good cross-section of typical maximum-minimum problems here.

13-1 The perimeter of a sector of a circle is 12 (units). Find the radius so that the area of the sector is a maximum.

Challenge Solve the problem for perimeter P.

13-2 The seating capacity of an auditorium is 600. For a certain performance, with the auditorium not filled to capacity, the receipts were $330.00. Admission prices were 75¢ for adults and 25¢ for children. If a represents the number of adults at the performance, find the minimum value of a satisfying the given conditions.

Challenge 1 Find the value of c for $a = 361$, where c represents the number of children.

Challenge 2 Find the value of c for $a = 360 + k, k = 1, 2, 3, \ldots$.

13-3 When the admission price to a ball game is 50 cents, 10,000 persons attend. For every increase of 5 cents in the admission price, 200 fewer (than the 10,000) attend. Find the admission price that yields the largest income.

Challenge 1 How many attend when the admission price is $1.50?

Challenge 2 Find the maximum income.

Challenge 3 Find the admission price yielding the largest income if, in addition to the conditions stated in the original problem, there is an additional expense of one dollar for every 100 persons in attendance.

13-4 A rectangle is inscribed in an isosceles triangle with base $2b$ (inches) and height h (inches), with one side of the rectangle lying in the base of the triangle. Let T (square inches) be the area of the triangle, and R_m the area of the largest rectangle so inscribed. Find the ratio $R_m:T$.

Challenge Change isosceles triangle to scalene triangle with base $2b$ and corresponding altitude h, and solve the problem.

13-5 It can be proved that the function $f(y) = ay - y^b$ (where $b > 1$, $a > 0$, and $y \geq 0$) takes its largest value when $y = \left(\frac{a}{b}\right)^{\frac{1}{b-1}}$. Use this theorem to find the maximum value of the function $F = \sin x \sin 2x$.

13-6 In the woods 12 miles north of a point B on an east-west road, a house is located at point A. A power line is to be built to A from a station at E on the road, 5 miles east of B. The line is to be built either directly from E to A or along the road to a point P (between E and B), and then through the woods from P to A, whichever is cheaper. If it costs twice as much per mile building through the woods as it does building along the highway, find the location of point P with respect to point B for the cheapest construction.

13-7 From a rectangular cardboard 12 by 14, an isosceles trapezoid and a square, of side length s, are removed so that their combined area is a maximum. Find the value of s.

Challenge Find the value of s for a combined area that is minimum.

13-8 Two equilateral triangles are to be constructed from a line segment of length L. Determine their perimeters P_1 and P_2 so that (a) the combined area is a maximum (b) the combined area is a minimum.

Challenge Verify that the maximum area is twice the minimum area.

13-9 Find the least value of $x^4 + y^4$ subject to the restriction $x^2 + y^2 = c^2$.

Challenge Find the least value of $x^3 + y^3$ subject to the restriction $x + y = c$.

13-10 Find the value of x such that $S = (x - k_1)^2 + (x - k_2)^2 + \cdots + (x - k_n)^2$ is a minimum where each k_i, $i = 1, 2, \ldots, n$, is a constant.

13-11 If $|x| \leq c$ and $|x - x_1| \leq 1$, find the greatest possible value of $|x_1^2 - x^2|$.

13-12 Show that the maximum value of $F = \dfrac{ab}{4(a + b)^2}$, where a, b are positive numbers, is $\dfrac{1}{16}$.

13-13 Find the area of the largest trapezoid that can be inscribed in a semicircle of radius r.

---

14 Quadratic Equations: Fair and Square

Two themes can be traced in this section. There are problems whose solution calls for solving a quadratic equation at some point. The other problems concern a variety of relationships between the roots (solutions) and coefficients of quadratic equations beyond the sum and product relationships usually studied in high school algebra.

14-1 Find the real values of x such that $3^{2x^2-7x+3} = 4^{x^2-x-6}$.

Challenge 1 Solve the problem for $3^{2x^2-7x+3} = 6^{x^2-x-6}$.

Challenge 2 Solve the problem for $4^{2x^2-7x+3} = 8^{x^2-x-6}$.

14-2 Let $D = h^2 + 3k^2 - 2hk$, where h, k are real numbers. For what values of h and k is $D > 0$?

Challenge If $D = h^2 - 3k^2 + 2hk$ with h and k real numbers, find the values of h and k for which **(a)** $D > 0$ **(b)** $D < 0$.

14-3 If the roots of $x^2 + bx + c = 0$ are the squares of the roots of $x^2 + x + 1 = 0$, find the values of b and c.

Challenge Solve the problem so that the roots of $x^2 + bx + c = 0$ are the cubes of the roots of $x^2 + x + 1 = 0$.

14-4 If the roots of $ax^2 + bx + c = 0$, $a \neq 0$, are in the ratio $m:n$, find an expression relating m and n to a, b, and c.

14-5 Find all values of x satisfying the pair of equations
$$x^2 - px + 20 = 0, \quad x^2 - 20x + p = 0.$$

14-6 A student, required to solve the equation $x^2 + bx + c = 0$, inadvertently solves the equation $x^2 + cx + b = 0$; b, c integers. One of the roots obtained is the same as a root of the original equation, but the second root is m less than the second root of the original equation. Find b and c in terms of m.

14-7 If r_1 and r_2 are the roots of $x^2 + bx + c = 0$, and $S_2 = r_1{}^2 + r_2{}^2$, $S_1 = r_1 + r_2$, and $S_0 = r_1{}^0 + r_2{}^0$, prove that $S_2 + bS_1 + cS_0 = 0$.

Challenge Find the relation between the roots r_1, r_2, and r_3 and the coefficients of a cubic equation $x^3 + bx^2 + cx + d = 0$, and then determine the value of $S_3 + bS_2 + cS_1 + dS_0$ where $S_3 = r_1{}^3 + r_2{}^3 + r_3{}^3$, $S_2 = r_1{}^2 + r_2{}^2 + r_3{}^2$, $S_1 = r_1 + r_2 + r_3$, $S_0 = r_1{}^0 + r_2{}^0 + r_3{}^0$.
HINT: Let $(x - r_1)(x - r_2)(x - r_3) \equiv x^3 + bx^2 + cx + d$.

14-8 A man sells a refrigerator for \$171, gaining on the sale as many percent (based on the cost) as the refrigerator cost, C, in dollars. Find C.

Challenge 1 Solve the problem if the percent gain is half the value of the new C.

Challenge 2 Solve the problem if the percent gain is half the value of the new C, and the selling price is \$170.50.

14-9 Express q and s each in terms of p and r so that the equation $x^4 + px^3 + qx^2 + rx + s = 0$ has two double roots u and v where u may or may not equal v. (Each of the factors $x - u$ and $x - v$ appears twice in the factorization of $x^4 + px^3 + qx^2 + rx + s$.)

Challenge 1   Find u, v if $p = -4$ and $r = -4$.

Challenge 2   Find u, v if $p = 2$ and $r = 2$.

14-10 Let $f(n) = n(n + 1)$ where n is a natural number. Find values of n such that $f(n + 4) = 4f(n) + 4$.

Challenge 1   Is there a pair m, n such that $2f(n) + 2 = f(m)$ where $m = n + 2$?

Challenge 2   Is there a pair m, n such that $2\left[\frac{1}{2}n(n + 1)\right] = \frac{1}{2}m(m + 1)$?

14-11 If one root of $Ax^3 + Bx^2 + Cx + D = 0$, $A \neq 0$, is the arithmetic mean of the other two roots, express the simplest relation between A, B, C, and D.

Challenge   Find the simplest relation between the coefficients if one root is the positive geometric mean of the other two.

14-12 If the coefficients a, b, c of the equation $ax^2 + bx + c = 0$ are odd integers, find a relation between a, b, c for which the roots are rational.

14-13 If $f(x) = a_0x^2 + a_1x + a_2 = 0$, $a_0 \neq 0$, and a_0, a_2 and $s = a_0 + a_1 + a_2$ are odd numbers, prove that $f(x) = 0$ has no rational root.

---

15 Systems of Equations: Strictly Simultaneous

Sometimes we can ask more interesting questions about a system of equations than "what is its solution." In other problems here, you will have your hands full hunting for the solution. It may be helpful to review Cramer's Rule before plunging in. (See Appendix VII.)

15-1 Estimate the values of the four variables in the given linear system. Then substitute repeatedly until a definitive solution is reached.

$$x_1 = \frac{1}{4}(0 + x_2 + x_3 + 0)$$
$$x_2 = \frac{1}{4}(0 + 0 + x_4 + x_1)$$
$$x_3 = \frac{1}{4}(x_1 + x_4 + 1 + 0)$$
$$x_4 = \frac{1}{4}(x_2 + 0 + 1 + x_3)$$

15-2 For the system
$$x + y + 2z = a$$
$$-2x - z = b$$
$$x + 3y + 5z = c$$
find a relation between a, b, and c so that a solution exists other than $x = 0$, $y = 0$, $z = 0$.

15-3 Find the smallest value of p^2 for which the pair of equations,
$$(4 - p^2)x + 2y = 0$$
$$2x + (7 - p^2)y = 0$$
has a solution other than $x = y = 0$, and find the ratio $x:y$ for this value of p^2.

15-4 If
$$P_1 = 2x^4 + 3x^3 - 4x^2 + 5x + 3,$$
$$P_2 = x^3 + 2x^2 - 3x + 1,$$
$$P_3 = x^4 + 2x^3 - x^2 + x + 2,$$
and $aP_1 + bP_2 + cP_3 = 0$, find the value of $a + b + c$, where $abc \neq 0$.

15-5 If
$$f_1 = 3x - y + 2z + w,$$
$$f_2 = 2x + 3y - z + 2w,$$
$$f_3 = 5x - 9y + 8z - w,$$
find numerical values of a, b, c so that $af_1 + bf_2 + cf_3 = 0$.

15-6 Find the common solutions of the set of equations
$$x - 2xy + 2y = -1$$
$$x - xy + y = 0.$$

Challenge Solve the problem replacing $x - xy + y = 0$ by $x - xy + y = 1$, and verify the result geometrically.

15-7 Solve the system $3x + 4y + 5z = a,$

$$4x + 5y + 6z = b,$$
$$5x + 6y + 7z = c,$$

a, b, c arbitrary real numbers, subject to the restriction $x \geq 0$, $y \geq 0$, $z > 0$.

15-8 For a class of N students, $15 < N < 30$, the following data were obtained from a test on which 65 or above is passing: the range of marks was from 30 to 90; the average for all was 66, the average for those passing was 71, and the average for those failing was 56. Based on a minor flaw in the wording of a problem, an upward adjustment of 5 points was made for all. Now the average mark of those passing became 79, and of those failing, 47. Find the number N_0 of students who passed originally, and the number N_f of those passing after adjustment, and N.

---

16   Algebra and Geometry: Often the Twain Shall Meet

Many mathematical ideas have both an algebraic aspect and a geometric aspect. Several of these ideas are explored here, with the emphasis on analytic geometry and transformations.

16-1 Curve I is the set of points (x, y) such that $x = u + 1$, $y = -2u + 3$, u a real number. Curve II is the set of points (x, y) such that $x = -2v + 2$, $y = 4v + 1$, v a real number. Find the number of common points.

Challenge 1   Change $y = -2u + 3$ to $y = -2u + 1$ and solve the problem.

Challenge 2   Change $x = u + 1$ to $x = 2u + 1$ and solve the problem.

16-2 Let the altitudes of equilateral triangle ABC be $\overline{AA}_1$, $\overline{BB}_1$, and $\overline{CC}_1$, with intersection point H. Let p represent a counterclockwise rotation of the triangle in its plane through 120° about point H. Let q represent a similar rotation through 240°. Let r represent a rotation of the triangle through 180° about line $\overleftrightarrow{AA_1}$. And let s, t represent similar rotations about lines $\overleftrightarrow{BB_1}$, $\overleftrightarrow{CC_1}$, respectively.

If we define $p * r$ to mean "first perform rotation p and then perform rotation r," find a simpler expression for $(q * r) * q$; that is, rotation q followed by rotation r, and this resulting rotation followed by rotation q.

Challenge 1 Determine whether these rotations are commutative; that is, for example, whether $p * r = r * p$.

Challenge 2 Determine whether these rotations are associative; that is, for example, whether $(q * r) * p = q * (r * p)$.

16-3 Fig. 16-3 represents a transformation of the segment $\overline{AB}$ onto segment $\overline{A'B'}$, and of $\overline{BC}$ onto $\overline{B'C'}$. The points of $\overline{AB}$ go into points of $\overline{A'B'}$ by parallel projections (parallel to $\overleftrightarrow{AA'}$). The points of $\overline{BC}$ go into points of $\overline{B'C'}$ by projections through the fixed point P.

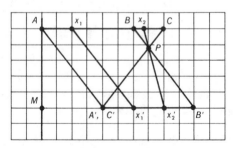

16-3

The distances from the left vertical line $\overline{AM}$ are zero for point A, 3 for point B, 4 for point C, 5 for point B', and 2 for point $A'(C')$. Designate the distances of the points on $\overline{AC}$ from $\overline{AM}$ as x, and the distances of their projections on $\overline{A'B'}(C'B')$ from $\overline{AM}$ as f. Find the values of r and s of the transformation functions $f = rx + s$ **(a)** for $0 \leq x \leq 3$ **(b)** for $3 \leq x \leq 4$.

16-4 Given the three equations (1) $7x - 12y = 42$ (2) $7x + 20y = 98$ (3) $21x + 12y = m$, find the value(s) of m for which the three lines form a triangle of zero area.

Challenge 1   Solve the problem generally for the system of equations (1) $a_1x + b_1y = c_1$ (2) $a_2x + b_2y = c_2$ (3) $a_3x + b_3y = m$ with $a_1b_2 - a_2b_1 \neq 0$.

Challenge 2   Solve the problem for (1) $4x - 6y = 21$, (2) $2x - 3y = 21$, (3) $21x + 12y = m$, and explain the "weird" result.

16-5 Describe the graph of $\sqrt{x^2 + y^2} = y$.

Challenge   Describe the graph of $\sqrt{x^2 + y^2} = -y$.

16-6 Transform $x^2 - 3x - 5 = 0$ into an equation of the form $aX^2 + b = 0$ where a and b are integers.

Challenge   Verify this transformation as a translation of the parabola $y = x^2 - 3x - 5$ in the xy-plane, a distance of $1\frac{1}{2}$ units to the left.

16-7 It is required to transform $2x_1{}^2 - 4x_1x_2 + 3x_2{}^2$ into an expression of the type $a_1y_1{}^2 + a_2y_2{}^2$. Using the transformation formulas $y_1 = x_1 + cx_2$ and $y_2 = x_2$, determine the values of a_1 and a_2.

Challenge 1   Investigate the case when the transformation formulas are $y_1 = x_1$ and $y_2 = dx_1 + x_2$.

Challenge 2   Investigate the case when the transformation formulas are $y_1 = x_1 + cx_2$ and $y_2 = dx_1 + x_2$.

16-8 *N.B.* and *S.B.* are, respectively, the north and south banks of a river with a uniform width of one mile. (See Fig. 16-8.) Town A is 3 miles north of *N.B.*, town B is 5 miles south of *S.B.* and 15 miles east of A. If crossing at the river banks is only at right angles to the banks, find the length of the shortest path from A to B.

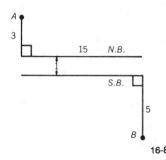

16-8

Challenge If the rate of land travel is uniformly 8 m.p.h., and the rowing rate on the river is $1\frac{2}{3}$ m.p h. (in still water) with a west to east current of $1\frac{1}{3}$ m.p.h., find the shortest time it takes to go from A to B.

16-9 Let the vertices of a triangle be $(0, 0)$, $(x, 0)$, and (hx, mx), m a positive constant and $0 \leq h < \infty$. Let a curve C be such that the y-coordinates of its points are numerically equal to the areas of the triangles for the values of h designated. Write the equation of curve C.

Challenge Let the vertices of a triangle be $(0, 0)$ and $(x, 0)$, and $\left(\frac{x}{2}, mx\right)$, m a positive constant. Let a curve C be such that the y-coordinates of its points are numerically equal to the perimeters of the triangles thus formed. Write the equation of curve C.

16-10 Each member of the family of parabolas $y = ax^2 + 2x + 3$ has a maximum or a minimum point dependent upon the value of a. Find an equation of the locus of the maxima and minima for all possible values of a.

16-11 Let A, B, and C be three distinct points in a plane such that $AB = X > 0$, $AC = 2AB$, and $AB + BC = AC + 2$. Find the values of X for which the three points may be the vertices of a triangle.

Challenge Find the value(s) of x for which the triangle is a right triangle.

16-12 The area of a given rectangle is 450 square inches. If the area remains the same when h inches are added to the width and h inches are subtracted from the length, find the new dimensions.

Challenge 1 Solve the problem when 12 inches are added to the width, and 10 inches subtracted from the length.

Challenge 2 Let the original width, W, be increased by h inches, and let the original length, L, be decreased by $\frac{1}{2}h$ inches. Express the new dimensions in terms of the original dimensions.

16-13 Show that if the lengths of the sides of a triangle are represented by a, b, and c, a necessary and sufficient condition for the triangle to be equilateral is the equality $a^2 + b^2 + c^2 = ab + bc + ca$. That is, if the triangle is equilateral, then $a^2 + b^2 + c^2 = ab + bc + ca$, and if $a^2 + b^2 + c^2 = ab + bc + ca$, then the triangle is equilateral.

16-14 Select point P in side $\overline{AB}$ of triangle ABC so that P is between A and the midpoint of $\overline{AB}$. Draw the polygon (not convex) $PP_1P_2P_3P_4P_5P_6$ such that $\overline{PP_1} \parallel \overline{AC}$, $\overline{P_1P_2} \parallel \overline{AB}$, $\overline{P_2P_3} \parallel \overline{CB}$, $\overline{P_3P_4} \parallel \overline{AC}$, $\overline{P_4P_5} \parallel \overline{AB}$, $\overline{P_5P_6} \parallel \overline{CB}$, with P_1, P_4 in $\overline{CB}$, P_2, P_5 in $\overline{AC}$, P_3, P_6 in $\overline{AB}$. Show that point P_6 coincides with point P.

Challenge 1   Investigate the case when P is the midpoint of $\overline{AB}$.

Challenge 2   Investigate the case when P is exterior to $\overline{AB}$ and $PA < AB$.

16-15 Let $P_1P_2P_3 \ldots P_nP_1$ be a regular n-gon (that is, an n-sided polygon) inscribed in a circle with radius 1 and center at the origin such that the coordinates of P_1 are $(1, 0)$. Let $S = (P_1P_2)^2 + (P_1P_3)^2 + \cdots + (P_1P_n)^2$. Find the ratio $S{:}n$.

17   Sequences and Series: Progression Procession

The problems in this section cover the territory between the usual study of sequences and series in high school and the calculus. The journey begins with unfamiliar facets of the familiar arithmetic and geometric sequences and ends with infinite series, touching on such things as recursive sequences, polynomial approximations, finite series, and limits along the way.

17-1 Find the last two digits of $N = 11^{10} - 1$.

Challenge 1   Find the last two digits of $N = 11^{10} + 1$.

Challenge 2 Find the last two digits of $N = 11^{10} - 9$.

17-2 Give a recursive definition of the sequence $\left\{\frac{1}{4n}\right\}$, n a natural number.

Challenge 1 Write a recursive definition of the sequence $\left\{\frac{1}{3n}\right\}$, $n = 1, 2, \ldots$.

Challenge 2 Write a recursive definition of the sequence $\left\{\frac{1}{3n-1}\right\}$, $n = 1, 2, \ldots$.

17-3 A perfectly elastic ball is dropped from height h feet. It strikes a perfectly elastic surface $\sqrt{\frac{h}{16}}$ seconds later. It rebounds to a height rh (feet), $0 < r < 1$, to begin a similar bounce a second time, then a third time, and so forth. Find **(a)** the total distance D (feet) traveled and **(b)** the total time T (seconds) to travel D feet.

Challenge From a pin O, two elastic spheres A and B are suspended with strings 1 foot long. Sphere A is brought to a horizontal position 1 foot from O and released. It strikes B, imparting motion to it, at the same time losing its own motion. B then falls and strikes A, imparting motion to it, at the same time losing its own motion. If with each impact the distance traveled is r times the preceding falling motion, $0 < r < 1$, find the total distances, D_A and D_B, traveled by A and B, respectively.

17-4 For the arithmetic sequence $a_1, a_2, \ldots, a_{16}$ it is known that $a_7 + a_9 = a_{16}$. Find each subsequence of three terms that forms a geometric sequence.

Challenge Find **(a)** each geometric subsequence of four terms, and **(b)** each geometric subsequence of five terms.

17-5 The sum of n terms of an arithmetic series is 216. The value of the first term is n and the value of the n-th term is $2n$. Find the common difference, d.

Challenge 1 Change the value of the n-th term to $11n$ and solve the problem.

Challenge 2 There is one more case where d is an integer. Find n and d.

17-6 Find the sum of n terms of the arithmetic series whose first term is the sum of the first n natural numbers and whose common difference is n.

Challenge Prove that the sum of n terms of the arithmetic series, whose first term is the sum of the first n odd natural numbers and whose common difference is n, is equal to the sum of n terms of the arithmetic series whose first term is the sum of the first n natural numbers and whose common difference is $2n$.

17-7 In a given arithmetic sequence the r-th term is s and the s-th term is r, $r \neq s$. Find the $(r + s)$-th term.

Challenge If $S_n = \frac{1}{2}(r + s)(r + s - 1)$, find n.

17-8 Define the triangular number T_n as $T_n = \frac{1}{2}n(n + 1)$, where $n = 0, 1, 2, \ldots, n, \ldots$ and the square number S_n as $S_n = n^2$, where $n = 0, 1, 2, \ldots, n, \ldots$. Prove **(a)** $T_{n+1} = T_n + n + 1$ **(b)** $S_{n+1} = S_n + 2n + 1$ **(c)** $S_{n+1} = T_{n+1} + T_n$ **(d)** $S_n = 2T_n - n$.

17-9 Beginning with the progression $a, ar, ar^2, ar^3, \ldots, ar^{n-1}, \ldots$, form a new progression by taking for its terms the differences of successive terms of the given progression, to wit, $ar - a$, $ar^2 - ar, \ldots$. Find the values of a and r for which the new progression is identical with the original.

17-10 The interior angles of a convex non-equiangular polygon of 9 sides are in arithmetic progression. Find the least positive integer that limits the upper value of the common difference between the measures of the angles.

Challenge 1 Find the least integer when there are 12 sides instead of 9.

Challenge 2 It is not much more difficult to solve the problem for the general case of n sides. Try it.

17-11 The division of $\frac{s}{s + r}$, where $r \ll s$, that is, where r is very much smaller in magnitude than s, is not exact, and is unending. If, however, we agree to stop at a given point, the quotient is a polynomial in $\frac{r}{s}$ whose degree depends upon the stopping point. Find a second-degree polynomial in $\frac{r}{s}$ best approximating the function $\frac{s}{s+r}$, $r \ll s$.

17-12 When $P_n(x) = 1 + x + x^2 + \cdots + x^n$ is used to approximate the function $P(x) = 1 + x + x^2 + \cdots + x^n + \cdots$ when $x = \frac{1}{4}$ (see Problem 9-11), find the smallest integer n such that $|P(x) - P_n(x)| < .001$.

Challenge 1   Solve the problem for $x = \frac{1}{2}$ and $x = \frac{1}{8}$.

Challenge 2   Solve the problem for $x = -\frac{1}{4}$, $x = -\frac{1}{2}$, and $x = -\frac{1}{8}$.

Challenge 3   Show that the values of n are large for those values of x that are toward the middle of the interval $(-1, +1)$.

17-13 Find the numerical value of S such that $S = a_0 + a_1 + a_2 + \cdots + a_n + \cdots$ where $a_0 = 1$, $a_n = r^n$, and $a_{n+2} = a_n - a_{n+1}$.

Challenge   Solve the problem with $a_0 = 2$.

17-14 A group of men working together at the same rate can finish a job in 45 hours. However, the men report to work singly at equal intervals over a period of time. Once on the job, however, each man stays until the job is finished. If the first man works five times as many hours as the last man, find the number of hours the first man works.

Challenge   What is the number of men?

17-15 A sequence of positive terms $A_1, A_2, \ldots, A_n, \ldots$ satisfies the recursive relation $A_{n+1} = \frac{3(1 + A_n)}{3 + A_n}$. For what values of A_1 is the sequence monotone decreasing (i.e., $A_1 \geq A_2 \geq \cdots \geq A_n \geq \cdots$)?

Challenge 1   Write several terms of the sequence when $A_1 = \sqrt{3}$.

Challenge 2   Write several terms of the sequence when $A_1 = 2$.

Challenge 3   Verify that the sequence is not monotone decreasing when $A_1 = 1\frac{1}{2}$.

17-16 If $S = n^3 + (n + 1)^3 + (n + 2)^3 + \cdots + (2n)^3$, n a positive integer, find S in closed form (that is, find a formula for S), given that $1^3 + 2^3 + \cdots + n^3 = \frac{1}{4}n^2(n + 1)^2$.

Challenge Find S where $S = 1^3 + 3^3 + 5^3 + \cdots + (2n - 1)^3$.

17-17 If $S(k) = 1 + 2 + 3 + \cdots + k$, express mn in terms of $S(m)$, $S(n)$, and $S(m + n)$.

Challenge Verify the formula for **(a)** $m = n$ **(b)** $m = 2n$ **(c)** $m = kn$.

17-18 Each a_i of the arithmetic sequence $a_0, a_1, 25, a_3, a_4$ is a positive integer. In the sequence there is a pair of consecutive terms whose squares differ by 399. Find the largest term of the sequence.

17-19 Let $S_1 = 1 + \cos^2 x + \cos^4 x + \cdots$; let $S_2 = 1 + \sin^2 x + \sin^4 x + \cdots$; let $S_3 = 1 + \sin^2 x \cos^2 x + \sin^4 x \cos^4 x + \cdots$, with $0 < x < \frac{\pi}{2}$. Show that $S_1 + S_2 = S_1 S_2$ and that $S_1 + S_2 + S_3 = S_1 S_2 S_3$.

17-20 A square array of natural numbers is formed as shown. Find the sum of the elements in **(a)** the j-th column **(b)** the i-th row **(c)** the principal diagonal (upper left corner to lower right corner).

1	2	3	.	.	n
$n + 1$	$n + 2$	.	.	.	$2n$
$2n + 1$	$2n + 2$	.	.	.	.
.	.	.	.	.	.
.	.	.	.	.	.
.	.	.	.	.	.
$(n - 1)n + 1$	$(n - 1)n + 2$	.	.	.	.

17-21 Let $S = 2x + 2x^3 + 2x^5 + \cdots + 2x^{2k-1} + \cdots$, where $|x| < 1$, be written as $\frac{1}{P} - \frac{1}{Q}$. Express P and Q as polynomials in x with integer coefficients.

Challenge 1 Evaluate S when $x = \frac{1}{2}$ and when $x = \frac{1}{4}$.

Challenge 2 Find P and Q if $S = 2 + 2x^2 + 2x^4 + \cdots + 2x^{2k} + \cdots$, where $|x| < 1$, and S is written as $\frac{1}{P} + \frac{1}{Q}$.

17-22 Let $I = \lim\limits_{n \to \infty} \dfrac{1^2 + 2^2 + \cdots + n^2}{n^3}$, that is, the limiting value of the fraction as n increases without bound; find the value of I.

17-23 Let $S = \dfrac{2}{1 \cdot 2} + \dfrac{2}{2 \cdot 3} + \dfrac{2}{3 \cdot 4} + \cdots + \dfrac{2}{n(n + 1)}$. Find a simple formula for S.

Challenge Find the values of $\lim\limits_{n \to \infty} S$; that is, $\dfrac{2}{1 \cdot 2} + \dfrac{2}{2 \cdot 3} + \dfrac{2}{3 \cdot 4} + \cdots$.

17-24 An endless series of rectangles is constructed on the curve $\dfrac{1}{x}$, each with width 1 and height $\dfrac{1}{n} - \dfrac{1}{n + 1}$, $n = 1, 2, 3, \ldots$. Find the total area of the rectangles.

17-25 Let $S_n = \dfrac{1}{1 \cdot 4} + \dfrac{1}{4 \cdot 7} + \cdots + \dfrac{1}{(3n - 2)(3n + 1)}$, where $n = 1$, $2, \ldots$. Find a simple formula for S_n in terms of n.

Challenge 1 Find $S_n = \dfrac{1}{1 \cdot 3} + \dfrac{1}{3 \cdot 5} + \cdots + \dfrac{1}{(2n - 1)(2n + 1)}$, where $n = 1, 2, \ldots$

Challenge 2 Can you now predict the formula for
$$S_n = \dfrac{1}{1 \cdot 2} + \dfrac{1}{2 \cdot 3} + \cdots + \dfrac{1}{n(n + 1)}, \text{where } n = 1, 2, \ldots ?$$

17-26 Let $S = a_1 + a_2 + \cdots + a_{n-1} + a_n$ be a geometric series with common ratio r, $r \neq 0$, $r \neq 1$. Let $T = b_1 + b_2 + \cdots + b_{n-1}$ be the series such that b_j is the arithmetic mean (average) of a_j and a_{j+1}, $j = 1, 2, 3, \ldots, n$. Express T in terms of a_1, a_n, and r.

17-27 The sum of a number and its reciprocal is 1. Find the sum of the n-th power of the number and the n-th power of the reciprocal.

17-28 Alpha travels uniformly 20 miles a day. Beta, starting from the same point three days later to overtake Alpha, travels at a uniform rate of 15 miles the first day, at a uniform rate of 19 miles the second day, and so forth in arithmetic progression. If n represents the number of days Alpha has traveled when Beta overtakes him, find n (not necessarily an integer).

17-29 Find a closed-form expression for S_n, where $S_n = 1 \cdot 2 + 2 \cdot 2^2 + 3 \cdot 2^3 + \cdots + n \cdot 2^n$; that is, find a simple formula for S.

Challenge 1   Find S_n where $S_n = 1 \cdot 3 + 2 \cdot 3^2 + 3 \cdot 3^3 + \cdots + n \cdot 3^n$.

Challenge 2   Find S_n where $S^n = 1 \cdot 4 + 2 \cdot 4^2 + 3 \cdot 4^3 + \cdots + n \cdot 4^n$.

Challenge 3   By inspection, find S_n where $S_n = 1 \cdot 5 + 2 \cdot 5^2 + 3 \cdot 5^3 + \cdots + n \cdot 5^n$.

17-30   Show that $\sum\limits_{r=1}^{\infty} \dfrac{1}{r^2} < 2$ where $\sum\limits_{r=1}^{\infty} \dfrac{1}{r^2} = \dfrac{1}{1^2} + \dfrac{1}{2^2} + \cdots + \dfrac{1}{r^2} + \cdots$.

17-31   Express S_n in terms of n, where $S_n = 1 \cdot 1! + 2 \cdot 2! + 3 \cdot 3! + \cdots + n \cdot n!$.

---

18 Logarithms: A Power Play

Logarithms were invented by John Napier in the early seventeenth century to simplify arithmetic computation. The advent of electronic computers in recent years has almost eliminated this practical need. But logarithms and logarithmic functions still have considerable theoretical importance. It's the theory that counts in these problems.

18-1   Find the real values of x such that $x \log_2 3 = \log_{10} 3$.

Challenge   Find the real values of y such that $y \log_{10} 3 = \log_2 3$.

18-2   Find the real values of x for which **(a)** F is real **(b)** F is positive, where $F = \log_a \dfrac{2x + 4}{3x}$, $a > 0$, $a \neq 1$.

Challenge   Change $2x + 4$ to $2x - 4$ and solve the problem.

18-3   If f is a function of x only and g is a function of y only, determine f and g such that $\log f + \log g = \log (1 + z)$ where $z = x + xy + y$.

Challenge   Solve the problem when $\log f + \log g = \log (1 - z)$ where $z = x - xy + y$.

18-4 If $(ax)^{\log a} = (bx)^{\log b}$, a, b positive, $a \neq b$, $a \neq 1$, $b \neq 1$, and the logarithmic base is the same throughout, express x in terms of a and b.

18-5 Find a simple formula for $S_n = \dfrac{1}{\log_2 N} + \dfrac{1}{\log_3 N} + \cdots + \dfrac{1}{\log_{25} N}$, $N > 1$.

Challenge Find a simple formula for $T_n = \dfrac{1}{\log_2 N} - \dfrac{1}{\log_3 N} + \dfrac{1}{\log_4 N} - \cdots - \dfrac{1}{\log_{25} N}$, $N > 1$.

---

19 Combinations and Probability: Choices and Chances

Handshakes at a party, a Ping-pong match, and a secret scientific project are among the settings for these problems, which involve (surprise!) counting choices or figuring chances.

19-1 Suppose that a boy remembers all but the last digit of his friend's telephone number. He decides to choose the last digit at random in an attempt to reach him. If he has only two dimes in his pocket (the price of a call is 10¢), find the probability that he dials the right number before running out of money.

Challenge Suppose that the boy remembers all but the last two digits, but he does know that their sum is 15. Find the probability of dialing correctly if only two dimes are available.

19-2 In a certain town there are 10,000 bicycles, each of which is assigned a license number from 1 to 10,000. No two bicycles have the same number. Find the probability that the number on the first bicycle one encounters will not have any 8's among its digits.

Challenge Find the probability that the number on the first bicycle one encounters will have neither an 8 nor a 7 among its digits.

19-3 Suppose Flash and Streak are equally strong Ping-pong players. Is it more probable that Flash will beat Streak in 3 games out of 4, or in 5 games out of 8?

Challenge   Suppose Flash is "twice as good" as Streak in the sense that, for many games, he wins twice as often as Streak. Is it more probable that Flash beats Streak in 3 games out of 5, or in 5 games out of 7?

19-4 Show that in a group of seven people it is impossible for each person to know reciprocally only three other persons.

Challenge   In a group of nine people, is it possible for each person to know reciprocally only five other persons?

19-5 At the conclusion of a party, a total of 28 handshakes was exchanged. Assuming that each guest was equally polite toward all the others, that is, each guest shook hands with each of the others, find the number of guests, n, at the party.

Challenge 1   Solve the problem for 36 handshakes.

Challenge 2   Solve the problem for 32 handshakes.

19-6 A section of a city is laid out in square blocks. In one direction the streets are E1, E2, ..., E7, and perpendicular to these are the streets N1, N2, ..., N6. Find the number of paths, each 11 blocks long, in going from the corner of E1 and N1 to the corner of E7 and N6.

19-7 A person, starting with 64 cents, makes 6 bets, winning three times and losing three times. The wins and losses come in random order, and each wager is for half the money remaining at the time the wager is made. If the chance for a win equals the chance for a loss, find the final result.

19-8 A committee of r people, planning a meeting, devise a method of telephoning s people each and asking each of these to telephone t new people. The method devised is such that no person is called more than once. Find the number of people, N, who are aware of the meeting.

19-9 Assume there are six line segments, three forming the sides of an equilateral triangle and the other three joining the vertices of the triangle to the center of the inscribed circle. It is required that the six segments be colored so that any two with a common point must have different colors. You may use any or all of 4 colors available. Find the number of different ways to do this.

19-10 A set of six points is such that each point is joined by either a blue string or a red string to each of the other five. Show that there exists at least one triangle completely blue or completely red. (See Fig. 19-10).

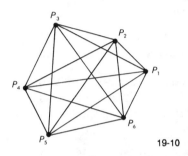

19-10

19-11 Each face of a cube is to be painted a different color, and six colors of paint are available. If two colorings are considered the same when one can be obtained from the other by rotating the cube, find the number of different ways the cube can be painted. [If the center of the cube is at the origin $(0, 0, 0)$ the rotations are about the x-axis, or the y-axis, or the z-axis through multiples of $90°$.]

19-12 An 8×8 checkerboard is placed with its corners at $(0, 0)$, $(8, 0)$, $(0, 8)$, and $(8, 8)$. Find the number of distinguishable non-square rectangles, with corners at points with integer coordinates, that can be counted on the checkerboard.

19-13 A group of 11 scientists are working on a secret project, the materials of which are kept in a safe. In order to permit the opening of the safe only when a majority of the group is present, the safe is provided with a number of different locks, and each scientist is given the keys to certain locks. Find the number of locks, n_1, required, and the number of keys, n_2, each scientist must have.

20 An Algebraic Potpourri

Here is an assortment of problems containing a few new ideas and a continuation of ideas from earlier sections.

20-1 If we define $(2n + 1)!$ to mean the product $(1)(2)(3) \ldots (2n + 1)$ and $(2n + 1)!!$ to mean the product $(1)(3)(5) \ldots (2n + 1)$, express $(2n + 1)!!$ in terms of $(2n + 1)!$.

20-2 If a, b, c are three consecutive odd integers such that $a < b < c$, find the value of $a^2 - 2b^2 + c^2$.

Challenge 1 How is the result changed if $c < b < a$?

Challenge 2 How is the result changed if three consecutive even integers are used?

20-3 At the endpoints A, B of a fixed segment of length L, lines are drawn meeting in C and making angles α, 2α, respectively, with the given segment. Let D be the foot of altitude $\overline{CD}$ and let x represent the length of $\overline{AD}$. Find the limiting value of x as α decreases towards zero; that is, find $\lim\limits_{\alpha \to 0} x$.

Challenge Note that, when the original angles are α, 2α, the limiting value of x is $\frac{2}{3}L$. Can you predict the limiting value of x when the original angles are α, 3α?

20-4 Find the set of integers $n \geq 1$ for which $\sqrt{n - 1} + \sqrt{n + 1}$ is rational.

Challenge 1 Solve the problem for $\sqrt{n - k} + \sqrt{n + k}$ where k is an integer such that $2 \leq k \leq 8$.

Challenge 2 For what positive integer values of n is $\sqrt{4n - 1}$ rational?

20-5 The angles of a triangle ABC are such that $\sin B + \sin C = 2 \sin A$. Find the value of $\tan \frac{B}{2} \tan \frac{C}{2}$.

Challenge Show that if $\sin B - \sin C = 2 \sin A$ for triangle ABC, then $\tan \frac{B}{2} \cot \frac{C}{2} = -3$.

20-6 Decompose $F = \dfrac{7x^3 - x^2 - x - 1}{x^3(x - 1)}$ into the sum of fractions with constant numerators.

Challenge Change the factor $x - 1$ in the denominator to $x + 1$ and solve the problem.

20-7 On a transcontinental airliner there are 9 boys, 5 American children, 9 men, 7 foreign boys, 14 Americans, 6 American males, and 7 foreign females. Find the number of people on the airliner.

20-8 If a and b are positive integers and b is not the square of an integer, find the relation between a and b so that the sum of $a + \sqrt{b}$ and its reciprocal is integral.

Challenge Solve the problem so that the sum is rational but not integral.

20-9 Find the simplest form for $R = \sqrt{1 + \sqrt{-3}} + \sqrt{1 - \sqrt{-3}}$.

Challenge 1 Simplify $S = \sqrt{1 + \sqrt{-3}} - \sqrt{1 - \sqrt{-3}}$.

Challenge 2 Simplify $T = \sqrt{a + \sqrt{-b}} \pm \sqrt{a - \sqrt{-b}}$, where a, b are positive integers.

20-10 Observe that the set $\{1, 2, 3, 4\}$ can be partitioned into subsets $T_1\{4, 1\}$ and $T_2\{3, 2\}$ so that the subsets have no element in common, and the sum of the elements in T_1 equals the sum of the elements in T_2. This cannot be done for the set $\{1, 2, 3, 4, 5\}$ or the set $\{1, 2, 3, 4, 5, 6\}$. For what values of n can a subset of the natural numbers $S_n = \{1, 2, 3, \ldots, n\}$ be so partitioned?

20-11 Suppose it is known that the weight of a medallion, X ounces, is represented by one of the integers $1, 2, 3, \ldots, N$. You have available a balance and two different weights, each with an integral number of ounces, represented by W_1 and W_2. Let $S = N + W_1 + W_2$. Find the value of S for the largest possible value of N that can be determined with the given conditions.

(For this problem we are indebted to Professor M. I. Aissen, Fordham University.)

20-12 If M is the midpoint of line segment $\overline{AB}$, and point P is between M and B, and point Q is beyond B such that $QP^2 = QA \cdot QB$, show that, with the proper choice of units, the length of $\overline{MP}$ equals the smaller root of $x^2 - 10x + 4 = 0$.

NOTE: A, M, B, P, and Q are collinear.

20-13 In Fig. 20-13, consider the lattice where R_i is the i-th row and C_j is the j-th column, $i, j, = 1, 2, 3, \ldots$, in which all the entries are natural numbers. Find the row and column for the entry 1036.

Challenge   Find the row and column for the entry 212.

20-13

20-14 Express $P(c) = c^6 + 10c^4 + 25c^2$ as a polynomial of least positive degree when c is a root of $x^3 + 3x^2 + 4 = 0$.

Challenge   By inspection solve the problem when $+25c^2$ is changed to $-25c^2$.

20-15 Let $S = \dfrac{b_0 + b_1x + \cdots + b_nx^n}{1 - x} = r_0 + r_1x + \cdots + r_nx^n$ be an identity in x. Express r_n in terms of the given b's.

Challenge   Find the coefficient of x^n when $S(x) = 1 + 2x + 3x^2 + \cdots + (n + 1)x^n$ is divided by $1 - x$.

20-16 Find the numerical value of the infinite product P whose factors are of the form $\dfrac{n^3 - 1}{n^3 + 1}$, where $n = 2, 3, 4, \ldots$.

20-17 Express $F = \dfrac{\sqrt[3]{2}}{1 + 5\sqrt[3]{2} + 7\sqrt[3]{4}}$ with a rational denominator.

Challenge   Express $F = \dfrac{\sqrt[3]{2}}{1 + \sqrt[3]{2} + \sqrt[3]{4}}$ with a rational denominator.

20-18 Starting with the line segment from 0 to 1 (including both endpoints), remove the open middle third; that is, points $\frac{1}{3}$ and $\frac{2}{3}$ of the middle third remain. Next remove the open middle thirds of the two remaining segments (points $\frac{1}{9}$, $\frac{2}{9}$, $\frac{7}{9}$, $\frac{8}{9}$ remain along with $\frac{1}{3}$ and $\frac{2}{3}$). Then remove the open middle thirds of the four segments remaining, and so on endlessly. Show that one of the remaining points is $\frac{1}{4}$.

20-19 Write a formula that can be used to calculate the n-th digit a_n of $N = .01001000100001 \ldots$, where all the digits are either 0 or 1, and where each succeeding block has one more zero than the previous block.

Challenge   Find a_n, the n-th digit, of $M = .101001000100001 \ldots$.

SOLUTIONS

1. Posers: Innocent and Sophisticated

1-1 *Suppose there are 6 pairs of blue socks all alike, and 6 pairs of black socks all alike, scrambled in a drawer. How many socks must be drawn out, all at once (in the dark), to be certain of getting a matching pair?*

Imagine two bins, the first marked "blue," the second marked "black." If the first sock drawn is blue, assign it to the first bin; if black, assign it to the second bin. Do the same with the second sock drawn. If one of the bins has two socks, you have a matching pair. If not, there is one sock in each bin. When a third sock is drawn, it must go in one or the other bin, thus giving a matching pair. Therefore, at most, 3 socks must be drawn out.

Challenge 1 *Suppose the drawer contains 3 black pairs of socks, 7 green pairs, and 4 blue pairs, scrambled. How many socks must be drawn out, all at once (in the dark), to be certain of getting a matching pair?*

Follow the pattern of reasoning in Problem 1-1. The answer is 4.

Challenge 2 *Suppose there are 6 different pairs of cuff links scrambled in a box. How many links must be drawn out, all at once (in the dark), to be certain of getting a matching pair?*

Follow the pattern of reasoning above. The answer is 7.

1-2 *Find five positive whole numbers a, b, c, d, e such that there is no subset with a sum divisible by 5.*

Difficult as this problem seems, it yields to an easy solution if you ask yourself the right questions!

What are the possible non-zero remainders when a positive integer is divided by 5? They are 1, 2, 3, 4, which you may think of as associated with bin 1, bin 2, bin 3, and bin 4, respectively. (See solution 1-1.)

Consider the five subsets $\{a\}$, $\{a, b\}$, $\{a, b, c\}$, $\{a, b, c, d\}$, $\{a, b, c, d, e\}$, and the respective sums of the elements in these subsets. When these sums are divided by 5, there are at most four different non-zero remainders. (Why are we not concerned with a zero remainder?) Therefore, at least two of these five sums, say s_2 and s_5, have the same remainder r, where r is 1 or 2 or 3 or 4, so that we may write $s_2 = 5m + r$ and $s_5 = 5n + r$, where m, n are integers. It follows that $s_5 - s_2 = 5m - 5n = 5p$, where $p = m - n$; that is, $s_5 - s_2$ is exactly divisible by 5. If $s_5 = a + b + c + d + e$, and $s_2 = a + b$, then $c + d + e$ is exactly divisible by 5. Hence, the sum of the elements in at least one subset of any five positive numbers is divisible by 5.

The method used for solving Problems 1-1, 1-2, 1-3, and 1-4 is aptly named the Pigeon-Hole Principle.

1-3 *A multiple dwelling has* 50 *letter boxes. If* 101 *pieces of mail are correctly delivered and boxed, show that there is at least one letter box with* 3 *or more pieces of mail.*

If 100 letters are distributed evenly among the 50 letter boxes, each box will contain 2 letters. When the 101st letter is put into a box, that box will contain 3 letters.

Challenge *What conclusion follows if there are*

    (a) 102 *pieces of mail*
    (b) 150 *pieces of mail*
    (c) 151 *pieces of mail?*

    (a) At least one box contains 3 or more letters.
    (b) At least one box contains 3 or more letters.
    (c) At least one box contains 4 or more letters.

1-4 *Assume that at least one of* a_1 *and* b_1 *has property* P, *and at least one of* a_2 *and* b_2 *has property* P, *and at least one of* a_3 *and* b_3 *has property* P. *Prove that at least two of* a_1, a_2, a_3, *or at least two of* b_1, b_2, b_3 *have property* P.

PROOF I: (Pigeon-Hole Principle) Use two pigeon holes or boxes labeled P and $\sim P$, where $\sim P$ means "not-P." The possibilities are as follows: Box P contains a_1 or b_1 or both, while box $\sim P$ contains a_1 or b_1 or neither. Similarly, box P contains a_2 or b_2 or both, and a_3 or b_3 or both. Box P, therefore, contains one of the following combinations:

$$a_1, a_2, a_3;\ a_1, a_2, b_3;\ a_1, a_2, a_3, b_3;\ a_1, b_2, a_3;\ a_1, a_2, b_2, b_3;$$

and so forth. Since P must contain at least three elements, at least two must be from the a's or at least two must be from the b's.

PROOF II: (Partitions) Assume that at most one of a_1, a_2, a_3, and at most one of b_1, b_2, b_3 have property P. Then, by simple addition, at most 2 of the 6 items have property P. This contradicts the given information which implies that at least 3 of the 6 items have property P. Hence, more than one of a_1, a_2, a_3, or more than one of b_1, b_2, b_3, have property P.

This reasoning is equivalent to the reasoning based on partitioning the numeral 3. Since $3 = 3 + 0 = 2 + 1 = 1 + 2 = 0 + 3$, we must have either 3 of the a's, or 2 of the a's and 1 of the b's, or 1 of the a's and 2 of the b's, or 3 of the b's with property P.

PROOF III: (Binomial Theorem) Let a_i, $i = 1, 2, 3$, represent one of the three given a's, and similarly for b_i. Since $(a_i + b_i)^3 = a_i^3 + 3a_i^2 b_i + 3a_i b_i^2 + b_i^3$ (see Appendix VI), then either the 3 a's or the 3 b's have property P, or 2 a's and 1 b, or 1 a and 2 b's have property P.

1-5 *An airplane flies round trip a distance of* L *miles each way. The velocity with head wind is* 160 *m.p.h., while the velocity with tail wind is* 240 *m.p.h. What is the average speed for the round trip?*

The temptation is to say that the average speed is $\frac{1}{2}(160 + 240) =$ 200 m.p.h. This is incorrect because the flying times are not the same for each leg of the trip.

Start with the basic formula $R \times T = D$, where R is the average speed (m.p.h.), T is the total time (hours), and D is the total distance (miles). Letting T_1 and T_2 be the times for the legs of the trip, we have $T_1 = \frac{L}{160}$ and $T_2 = \frac{L}{240}$.

Therefore, $T = T_1 + T_2 = \dfrac{L}{160} + \dfrac{L}{240} = L\left(\dfrac{1}{96}\right)$. Since the total distance is $2L$, $R = \dfrac{2L}{L\left(\dfrac{1}{96}\right)} = 192$ m.p.h. The result is the Harmonic Mean between the two velocities. (See Appendix IV.) (i.e., here H.M. $= \dfrac{2(160)(240)}{160 + 240} = 192$ m.p.h.)

1-6 *Assume that the trains between New York and Washington leave each city every hour on the hour. On its run from Washington to New York, a train will meet n trains going in the opposite direction. If the one-way trip in either direction requires four hours exactly, what is the value of n?*

At the moment the train leaves Washington, say 2 P.M., the train that left New York at 10 A.M. is pulling into the station. When it reaches New York at 6 P.M., a train is leaving New York for Washington. In all, it meets 9 trains coming from New York; to wit, those that left New York at 10, 11, 12, 1, 2, 3, 4, 5, and 6.

1-7 *A freight train one mile long is traveling at a steady speed of 20 miles per hour. It enters a tunnel one mile long at 1 P.M. At what time does the rear of the train emerge from the tunnel?*

To clear the tunnel, the train must travel a distance of 2 miles, the length of the tunnel plus the length of the train. At 20 m.p.h., the train travels 1 mile in 3 minutes, and 2 miles in 6 minutes. The rear of the train emerges at 1:06 P.M.

1-8 *A watch is stopped for 15 minutes every hour on the hour. How many actual hours elapse during the interval the watch shows 12 noon to 12 midnight?*

The total elapsed time is 12 plus the number of hours lost; that is, $12 + 11\left(\dfrac{1}{4}\right) = 14\dfrac{3}{4}$ (hours).

Challenge 2 *Between 12 noon and 12 midnight, a watch is stopped for 1 minute at the end of the first full hour, for 2 minutes at the end of the second full hour, for 3 minutes at the end of the third full hour, and so forth for the remaining full hours. What is the true time when this watch shows 12 midnight?*

The total elapsed time, in hours, is $12 + \frac{1}{60}\left(\frac{1}{2} \cdot 11 \cdot 12\right) = 13\frac{1}{10}$. (See Appendix VII.) Therefore, the true time is 1:06 A.M.

1-9 *The last three digits of a number* N *are* x25. *For how many values of* x *can* N *be the square of an integer?*

Let M be such that $M^2 = N$. Since the last two digits of N are 25, M terminates in 5. If $M = 5$, $x = 0$.

If $x \neq 0$, then M has at least two digits. When the tens' digit of M is 1, 3, 6, or 8, then $x = 2$. When the tens' digit of M is 2 or 7, then $x = 6$. When the tens' digit of M is 4, 5, or 9, then $x = 0$. In all, there are three possible values for x, namely 0, 2, or 6.

1-10 *A man born in the eighteenth century was* x *years old in the year* x^2. *How old was he in 1776? (Make no correction for calendric changes.)*

In the interval 1700-1800, the only square of an integer is $1764 = 42^2$. Therefore, the year of his birth was $1764 - 42 = 1722$. Hence, in 1776, he was $1776 - 1722 = 54$ years old.

1-11 *To conserve the contents of a* 16 *oz. bottle of tonic, a castaway adopts the following procedure. On the first day he drinks* 1 *oz. of tonic, and then refills the bottle with water; on the second day he drinks* 2 *oz. of the mixture, and then refills the bottle with water; on the third day he drinks* 3 *oz. of the mixture, and again refills the bottle with water. The procedure is continued for succeeding days until the bottle is empty. How many ounces of water does he thus drink?*

It is very easy to get bogged down in a problem like this. You must resist the urge to find the daily ratios of tonic to water, or of water to mixture. These are difficult to find, and unnecessary!

The essential clue is in the total number of ounces of water added during the drinking period.

On the first day, 1 oz. of water is added, on the fifteenth day, 15 oz. of water are added. (Explain why none was added on the sixteenth day.) The total is, therefore, $1 + 2 + \cdots + 15 = \frac{1}{2}(15)(16) = 120$ oz. (See Appendix VII.)

1-12 *Which yields a larger amount with the same starting salary:*
Plan I, with four annual increases of $100 each, or
Plan II, with two biennial increases of $200 each?

Let A (dollars) be the starting salary. Under Plan I the earnings by the end of the fifth year are

$$A + (A + 100) + (A + 200) + (A + 300) + (A + 400)$$
$$= 5A + 1000.$$

Under Plan II the earnings for the same period are

$$A + A + (A + 200) + (A + 200) + (A + 400) = 5A + 800.$$

Plan I yields $200 more.

1-13 *Assuming that in a group of* n *people any acquaintances are mutual, prove that there are two persons with the same number of acquaintances.*

PROOF: For every person in the group, the number of acquaintances is either 0, or 1, or 2, or . . . , or $n - 1$. Let us first assume that no two persons have the same number of acquaintances, so that each of the n numbers, 0 to $n - 1$, is represented. But since the presence of 0 means that there is a person acquainted with *no one*, and the presence of $n - 1$ means that there is a person acquainted with *everyone*, then our assumption that no two persons have the same number of acquaintances leads to a contradiction. This assumption is untenable. We are forced to conclude that there are two persons with the same number of acquaintances.

1-14 *The smallest of* n *consecutive integers is* j. *Represent in terms of* j
(a) *the largest integer* L (b) *the middle integer* M.

(a) $L = j + n - 1$ (b) $M = j + \dfrac{n - 1}{2}$, if n is odd. If n is even, there is no unique middle integer. We can designate either $j + \dfrac{n}{2} - 1$ or $j + \dfrac{n}{2}$ as middle integers, if n is even.

1-15 *We define the symbol* $|x|$ *to mean the value* x *if* $x \geq 0$, *and the value* $-x$ *if* $x < 0$. *Express* $|x - y|$ *in terms of* max(x, y) *and* min(x, y) *where* max(x, y) *means* x *if* $x > y$, *and* y *if* $x < y$, *and* min(x, y) *means* x *if* $x < y$, *and* y *if* $x > y$.

If $x > y$, $|x - y| = x - y = \max(x, y) - \min(x, y)$.
If $x < y$, $|x - y| = y - x = \max(x, y) - \min(x, y)$.
Therefore, $|x - y| = \max(x, y) - \min(x, y)$ in all cases.

1-16 *Let* $x^+ = \begin{cases} x \ if \ x \geq 0 \\ 0 \ if \ x < 0, \end{cases}$ *and let* $x^- = \begin{cases} -x \ if \ x \leq 0 \\ 0 \ if \ x > 0. \end{cases}$

Express:

(a) x *in terms of* x^+ *and* x^- (b) $|x|$ *in terms of* x^+ *and* x^-
(c) x^+ *in terms of* $|x|$ *and* x (d) x^- *in terms of* $|x|$ *and* x.

(a) If $x < 0$, then $x^+ = 0$ and $x^- = -x$, $\therefore x = x^+ - x^-$.
    If $x > 0$, then $x^+ = x$ and $x^- = 0$, $\therefore x = x^+ - x^-$.
    If $x = 0$, then $x^+ = x$ and $x^- = -x$, $\therefore x + x = x = x^+ - x^-$.

(b) If $x < 0$, then $x^+ = 0$ and $x^- = -x$ and $|x| = -x$,
    $\therefore |x| = x^+ + x^-$.
    If $x > 0$, then $x^+ = x$ and $x^- = 0$ and $|x| = x$, $\therefore |x| = x^+ + x^-$.
    If $x = 0$, then $x^+ = x$ and $x^- = -x$ and $|x| = x$, $\therefore |x| = x^+ + x^-$.

(c) If $x < 0$, then $x^+ = 0$. Since $\frac{1}{2}(|x| + x) = \frac{1}{2}(-x + x) = 0$,
    $x^+ = \frac{1}{2}(|x| + x)$.
    If $x > 0$, then $x^+ = x$. Since $\frac{1}{2}(|x| + x) = \frac{1}{2}(x + x) = x$,
    $x^+ = \frac{1}{2}(|x| + x)$.
    If $x = 0$, then $x^+ = x$. Since $\frac{1}{2}(|x| + x) = \frac{1}{2}(x + x) = x$,
    $x^+ = \frac{1}{2}(|x| + x)$.

(d) If $x < 0$, then $x^- = -x$. Since $\frac{1}{2}(|x| - x) = \frac{1}{2}(-x - x) = -x$, $x^- = \frac{1}{2}(|x| - x)$.
    If $x > 0$, then $x^- = 0$. Since $\frac{1}{2}(|x| - x) = \frac{1}{2}(x - x) = 0$,
    $x^- = \frac{1}{2}(|x| - x)$.
    If $x = 0$, then $x^- = -x$. Since $\frac{1}{2}(|x| - x) = \frac{1}{2}(-x - x) = -x$, $x^- = \frac{1}{2}(|x| - x)$.

1-17 *We define the symbol* [x] *to mean the greatest integer which is not greater than* x *itself. Find the value of* [y] + [1 − y].

(a) If y is an integer then $[y] + [1 - y] = y + 1 - y = 1$.

(b) If y is not an integer, let $y = a + h$ where $0 < h < 1$, and a is an integer. Then $[y] + [1 - y] = [a + h] + [1 - (a + h)] = a - a = 0$. Note that $[1 - (a + h)] = [-a + (1 - h)] = -a$.

ILLUSTRATIONS: For $y = 6$, $[y] + [1 - y] = 6 + 1 - 6 = 1$.
For $y = 6.37$, $[y] + [1 - y] = [6.37] + [1 - 6.37] = 6 - 6 = 0$.
For $y = -6$, $[y] + [1 - y] = [-6] + [7] = -6 + 7 = 1$.
For $y = -6.37$, $[y] + [1 - y] = [-6.37] + [7.37] = -7 + 7 = 0$.

Challenge 5 *Let* $(x) = x - [x]$; *express* $(x + y)$ *in terms of* (x) *and* (y).

Let $x = a + \epsilon_1$ where $0 \le \epsilon_1 < 1$, and let $y = b + \epsilon_2$ where $0 \le \epsilon_2 < 1$, where a and b are integers. Then $(x) = a + \epsilon_1 - a = \epsilon_1$, and $(y) = b + \epsilon_2 - b = \epsilon_2$. $x + y = a + \epsilon_1 + b + \epsilon_2$. If $\epsilon_1 + \epsilon_2 < 1$, then $(x + y) = a + b + \epsilon_1 + \epsilon_2 - a - b = \epsilon_1 + \epsilon_2$. Therefore, $(x + y) = (x) + (y)$ when $\epsilon_1 + \epsilon_2 < 1$. If $\epsilon_1 + \epsilon_2 \ge 1$, then $x + y = a + b + 1 + \epsilon_3$, where $\epsilon_3 + 1 = \epsilon_2 + \epsilon_1$ and $0 \le \epsilon_3 < 1$, since $\epsilon_1 + \epsilon_2 < 2$. Then $(x + y) = a + b + 1 + \epsilon_3 - a - b - 1 = \epsilon_3$. Therefore, $(x + y) = (x) + (y) - 1$ when $\epsilon_1 + \epsilon_2 \ge 1$.

1-18 *At what time after* 4:00 *will the minute hand overtake the hour hand?*

We may deal with this problem as we would if asked to find the time it took a fast car to overtake a slower one. Let us speak of the rate (i.e., speed) of the hour hand as r. Then the rate of the minute hand is $12r$.

The distance that the hands travel will be measured by the minute markers of the clock. What we seek in this problem is the distance (in minutes) that the minute hand must travel to overtake the hour hand. Let this distance be x. Therefore, the distance that the hour hand must travel is $x - 20$, since it has a 20-minute head start over the minute hand.

Since the time equals the distance divided by the rate, the time that the minute hand travels is $\frac{x}{12r}$ and the time that the hour hand travels is $\frac{x - 20}{r}$. Since both hands travel the same amount of time:

$$\frac{x}{12r} = \frac{x - 20}{r}; \quad x = \frac{12}{11} \cdot 20; \quad x = 21\frac{9}{11}.$$

Therefore, at $4:21\frac{9}{11}$ the minute hand will overtake the hour hand.

NOTE: In the relation, $x = \frac{12}{11} \cdot 20$, the quantity 20 is the number of minutes of head start that the hour hand has over the minute hand. The ratio $\frac{12}{11}$, then, is the number of minutes required, per minute of head start, for the minute hand to overtake the hour hand. Therefore, we can substitute any known value for the 20, and find the time required by multiplying by $\frac{12}{11}$.

For example, if the time at the start is 8:00, $x = \frac{12}{11} \cdot 40 = 43\frac{7}{11}$. The minute hand will overtake the hour hand at $8:43\frac{7}{11}$.

Another way of looking at this relation is to say that the head start of 20 minutes (or whatever the amount) is the time it would take the minute hand to reach the hour hand if the hour hand did not move. But because the hour hand does move, it takes $\frac{12}{11}$ as long. This idea can be applied to many variations of the clock problem.

Challenge 1   *At what time after* 7:30 *will the hands of a clock be perpendicular?*

Let us assume that the hour hand does not move after 7:00. Then the minute hand would be perpendicular to the hour hand at 7:50. (This would happen at 7:20 also, but the problem asks for a time after 7:30.) The travel time for the minute hand, with a stationary hour hand, is 50 minutes. With a moving hour hand, it must be $\frac{12}{11}$ as much: $\frac{12}{11} \cdot 50 = 54\frac{6}{11}$.

The hands will be perpendicular at $7:54\frac{6}{11}$.

Challenge 2   *Between* 3:00 *and* 4:00, *Noreen looked at her watch and noticed that the minute hand was between 5 and 6. Later, Noreen looked again and noticed that the hour hand and the minute hand had exchanged places. What time was it in the second case?*

Let $x =$ position of minute hand between 3:00 and 4:00; then the position of the hour hand is $\left(15 + \frac{x}{12}\right)$. Let

y = position of minute hand between 5:00 and 6:00; then the position of the hour hand is $\left(25 + \frac{x}{12}\right)$. Since the hands have changed places, $15 + \frac{x}{12} = y$ and $25 + \frac{y}{12} = x$. Solve for y; $y = 17\frac{29}{143}$.

ANSWER: $5:17\frac{29}{143}$

Challenge 3 *The hands of Ernie's clock overlap exactly every 65 minutes. If, according to Ernie's clock, he begins working at 9 A.M. and finishes at 5 P.M., how long does Ernie work according to an accurate clock?*

Using the technique described in Problem 1-18, we find that the hands of an accurate clock overlap every $65\frac{5}{11}$ minutes. Therefore, we may employ the following proportion:

$$\frac{65 \text{ minutes}}{65\frac{5}{11} \text{ minutes}} = \frac{8 \text{ hours}}{h \text{ hours}}, \quad h = 8\frac{8}{143}.$$

2 Arithmetic: Mean and Otherwise

2-1 *The arithmetic mean (A.M.), or ordinary average, of a set of 50 numbers is 32. The A.M. of a second set of 70 numbers is 53. Find the A.M. of the numbers in the sets combined.*

One procedure is to add the 50 numbers of the first set, add the 70 numbers of the second set, add these two sums, and then divide by 120. But, since the individual numbers are not known, we cannot use this method.

Nevertheless, we can obtain the essential information needed for a solution without a knowledge of the individual numbers. Since A.M. of n numbers $= \frac{S}{n}$, where S is the sum of the n numbers, $S = n(\text{A.M.})$. For the first set of numbers, $S_1 = 50 \times$

32, and for the second set of numbers, $S_2 = 70 \times 53$. Therefore, the required A.M. is $\dfrac{(50)(32) + (70)(53)}{50 + 70} = 44.25$.

Challenge 1 *Change the A.M. of the second set to* -53, *and solve.*

$$\text{A.M.} = \frac{(50)(32) + 70(-53)}{50 + 70} = -\frac{2110}{120} = -17\frac{7}{12}$$

Challenge 2 *Change the number of elements in each set to* 1, *and solve.*

$$\text{A.M.} = \frac{(1)(32) + (1 \times 53)}{1 + 1} = \frac{32 + 53}{2} = 42.5$$

Would you conclude from this illustration that finding the average of two numbers in the usual manner is a special case of this method, known as the method of Weighted Means?

A general representation of the method of Weighted Means may be given as

$$\text{A.M.} = \frac{(P_1)(M_1) + (P_2)(M_2) + \cdots + (P_n)(M_n)}{P_1 + P_2 + \cdots + P_n},$$

where M_i is the arithmetic mean of the set of numbers with P_i members in it, i being used for any one of the natural numbers $1, 2, 3, \ldots, n$. There are also other interpretations of the formula.

Challenge 3 *Find the point-average of a student with A in mathematics, A in physics, B in chemistry, B in English, and C in history — using the scale: A, 5 points; B, 4 points; C, 3 points; D, 1 point — when* **(a)** *the credits for the courses are equal* **(b)** *the credits for the courses are mathematics, 4; physics, 4; chemistry, 3; English, 3; and history, 3.*

$$\text{A.M.} = \frac{(2)(5) + (2)(4) + (1)(3)}{2 + 2 + 1} = 4.2, \text{ when the courses}$$
carry equal credit.

$$\text{A.M.} = \frac{(2)(4)(5) + (2)(3)(4) + (1)(3)(3)}{(2)(4) + (2)(3) + (1)(3)} = 4.3, \text{ when the}$$
credits are, respectively, 4, 4, 3, 3, 3.

Challenge 7 *Estimate the approximate A.M. of the set* $\{61, 62, 63, 65, 68, 73, 81, 94\}$.

When the numbers of a set are large, and addition becomes cumbersome, we frequently use the method of Guessed Mean to find the A.M. of a set of numbers.

For example, find the A.M. of the set $\{61, 62, 63, 65,$ $68, 73, 81, 94\}$. Obviously, the A.M. is between the extremes 61 and 94. Let us guess 70 as the A.M. (Would 80 be a good guess?) The differences between the given numbers and the guessed mean are $-9, -8, -7, -5,$ $-2, +3, +11, +24$. Their sum is $+7$. The true A.M. is, therefore, $70 + \frac{7}{8} = 70\frac{7}{8}$.

In rare instances the guessed mean will be the correct A.M. What is the sum of the differences in this instance? ANSWER: Zero

2-2 *Express the difference of the squares of two consecutive even integers in terms of their arithmetic mean.*

$|(2n + 2)^2 - (2n)^2| = |(4n + 2)(2)| = |4(2n + 1)| = 4|\text{A.M.}|$ where $|x|$ means $+x$ if $x \geq 0$, and $-x$ if $x < 0$.

2-3 *It is a fundamental theorem in arithmetic that a natural number can be factored into prime factors in only one way — if the order in which the factors are written is ignored. This is known as the Unique Factorization Theorem. For example, 12 is uniquely factored into the primes 2, 2, 3.*

Consider the set $S_1 = \{4, 7, 10, \ldots, 3k + 1, \ldots\}$, in which $k = 1, 2, \ldots, n, \ldots$. Does S_1 have unique factorization?

First, it is important to identify some of the prime members of S_1; that is, those members divisible by themselves but no other members of S_1. These are (with some surprises) 4, 7, 10, 13, 19, 22, 25, ..., since each of these numbers is (exactly) divisible only by itself.

After a few trials you find that $4 \times 25 = 10 \times 10 = 100$, so that 100 has two different factorizations in S_1. Another instance is $484 = 22 \times 22 = 4 \times 121$. Hence, S_1 does not have the property of unique factorization.

Challenge *Is factorization unique in $S_2 = \{3, 4, 5, \ldots, k, \ldots\}$?*

Of course, 3, 5, 7, 11, ... are primes in S_2, but so are 4, 6, 8, 10, ... Factorization in S_2 is not unique since $24 = 4 \times 6 = 3 \times 8$. Can you find another instance? ANSWER: $36 = 6 \times 6 = 3 \times 3 \times 4$

2-4 *What is the smallest positive value of* n *for which* $n^2 + n + 41$ *is not a prime number?*

It may seem strange, but the expression $n^2 + n + 41$ does generate prime numbers for every natural number n from 1 to 39. However, since $n^2 + n = n(n + 1)$, it cannot be the case that $n^2 + n + 41$ is prime when $n + 1 = 41$; that is, when $n = 40$ because, then, we have $40(41) + 41 = 41^2$, divisible by 41.

COMMENT: $n^2 + n + 41$ generates the same set of primes for the negative integers -1 to -40, but for $n = -41$, the expression is composite, not prime.

2-5 *Given the positive integers* a, b, c, d *with* $\frac{a}{b} < \frac{c}{d} < 1$; *arrange in order of increasing magnitude the five quantities:* $\frac{b}{a}, \frac{d}{c}, \frac{bd}{ac}, \frac{b+d}{a+c}, 1$.

Since $\frac{c}{d} < 1$, $\frac{d}{c} > 1$. Since $\frac{a}{b} < \frac{c}{d}$, $\frac{b}{a} > \frac{d}{c}$. Hence, $1 < \frac{d}{c} < \frac{b}{a}$.

Since each of $\frac{d}{c}$ and $\frac{b}{a}$ is greater than 1, then $\frac{bd}{ac}$ is greater than either. Hence, $1 < \frac{d}{c} < \frac{b}{a} < \frac{bd}{ac}$.

We now show that the fraction $\frac{b+d}{a+c}$, obtained by adding separately the numerators and the denominators of $\frac{b}{a}$ and $\frac{d}{c}$, is greater than the smaller fraction $\frac{d}{c}$, and less than the larger fraction $\frac{b}{a}$.

Since $\frac{d}{c} < \frac{b}{a}$, $ad < bc$; therefore, $cd + ad < cd + bc$, and $d(a + c) < c(d + b)$. $\frac{d}{c} < \frac{b+d}{a+c}$. Also, since $bc > ad$, $ab + bc > ab + ad$, $b(a + c) > a(b + d)$, and $\therefore \frac{b+d}{a+c} < \frac{b}{a}$.

The required order is, therefore, $1, \frac{d}{c}, \frac{b+d}{a+c}, \frac{b}{a}, \frac{bd}{ac}$.

2-6 *It can be proved (see Appendix I) that, for any natural number* n, *the terminal digit of* n^5 *is the same as that of* n *itself; that is,* n^5TD n, *where the symbol* TD *means "has the same terminal digit."* *For example,* 4^5 TD 4.

Find the terminal digit of **(a)** 2^{12} **(b)** 2^{30} **(c)** 7^7 **(d)** 8^{10} **(e)** $8^{10} \cdot 7''$

(a) Since $12 = 5 + 5 + 2$, $2^{12} = 2^5 \cdot 2^5 \cdot 2^2$ TD $2 \cdot 2 \cdot 2^2 = 16$ TD 6. (See Appendix I.) **(b)** $2^{30} = (2^5)^6$ TD $2^6 = 2^5 \cdot 2$ TD $2 \cdot 2 = 4$. **(c)** $7^7 = 7^5 \cdot 7^2$ TD $7 \cdot 7^2 = 343$ TD 3.

Challenge *Find the terminating digit of* (a) $\left(\frac{5}{8}\right)^5$ (b) $\left(\frac{4}{7}\right)^5$.

METHOD I: (a) $\left(\frac{5}{8}\right)^5 = \frac{5^5}{8^5}$ TD $\frac{5}{8}$. But $\frac{5}{8} = .625$, so that the terminating digit is 5.

METHOD II: (a) $\left(\frac{5}{8}\right)^5 = (.625)^5 = (625 \times 10^{-3})^5 = 625^5 \times 10^{-15}$. But 625^5 TD 5, so that the terminating digit of $\left(\frac{5}{8}\right)^5$ is 5. (b) $\left(\frac{4}{7}\right)^5 = (.\overline{571428})^5$; that is, the fifth power of an endless repeating decimal with period 571428. Since $(571428)^5$ TD 8, the terminating digit is 8, provided the complete block of digits 571428 is used.

2-7 *If* N $= 1 \cdot 2 \cdot 3 \cdots 100$ (*more conveniently written* 100!), *find the number of terminating zeros when the multiplications are carried out.*

For each factor 10 in *N* there will be a terminating zero; that is, for every pair of factors 5 and 2, there will be a terminating zero. We must, therefore, find the number of factors 5 and the number of factors 2 in 100!

The factor 5 is present once in each of 5, 10, 15, 20, 30, . . . , 95 (16 factors), and twice in each of 25, 50, 75, and 100 (8 factors); a total of 24 factors. The factor 2 is present once in each of 2, 6, 10, 14, . . . , 98, twice in each of 4, 12, 20, 28, . . . , 100, three times in each of 8, 24, . . . , 88, and so forth. (Show that there is a total of 97 factors 2.) Of these, only 24 are needed to pair with the 24 factors 5. The number of terminating zeros in 100! is, 24.

More elegantly, we find the number of factors 5 as follows: $100 \div 5 = 20$; $20 \div 5 = 4$; $20 + 4 = 24$. Similarly, for the factor 2, we find 97 factors. (The remainders in the divisions are disregarded; explain why.)

2-8 *Find the maximum value of* x *such that* 2^x *divides* 21!

METHOD I: List all the even factors in 21!, namely, 2, 4, . . . , 20, and the highest power of 2 in each, to obtain $1 + 2 + 1 + 3 + 1 + 2 + 1 + 4 + 1 + 2 = 18$. . . x(max) $= 18$.

METHOD II: Divide 2 into 21 and into the successive quotients until a quotient less than 2 is obtained. Then add the quotients.

$21 \div 2$ gives $q_1 = 10$; $10 \div 2$ gives $q_2 = 5$; $5 \div 2$ gives $q_3 = 2$; $2 \div 2$ gives $q_4 = 1$. $q_1 + q_2 + q_3 + q_4 = 10 + 5 + 2 + 1 = 18$.

Challenge 2   *Find the highest power of 2 in 21! excluding factors also divisible by 3.*

Factors divisible by 2 and 3 are $6 = 2 \cdot 3$, $12 = 2^2 \cdot 3$, $18 = 2 \cdot 3^2$. Therefore, the highest power is $18 - 4 = 14$.

2-9 *The number 1234 is not divisible by 11, but the number 1243, obtained by rearranging the digits, is divisible by 11. Find all the rearrangements that are divisible by 11.*

For an integer to be divisible by 11, the sum of the odd-numbered digits minus the sum of the even-numbered digits, must equal a multiple of 11. (See Appendix V.)

Since $2 + 3 = 1 + 4$, all rearrangements in which the odd-numbered digits are 2 and 3, or in which the even-numbered digits are 2 and 3, are acceptable.

In all, there are 8 rearrangements: 2134, 2431, 3124, 3421, 1243, 4213, 1342, 4312.

Challenge   *Solve the problem for 12034.*

All rearrangements in which the odd-numbered digits are 1, 0, 4, or in which the odd-numbered digits are 2, 0, 3; eight rearrangements in all.

2-10 *Let k be the number of positive integers that leave a remainder of 24 when divided into 4049. Find k.*

By definition, Dividend = Quotient $\times$ Divisor $+$ Remainder; or, stated in symbols, $D = qd + r$, where $0 \leq r < d$. Therefore, $qd = D - r = 4049 - 24 = 4025 = 5 \cdot 5 \cdot 7 \cdot 23$.

There are, therefore, one divisor using all four prime factors $(5 \cdot 5 \cdot 7 \cdot 23)$, three divisors using three of the prime factors at a time $(5 \cdot 5 \cdot 7, 5 \cdot 5 \cdot 23, 5 \cdot 7 \cdot 23)$, four divisors using two of the prime factors at a time $(5 \cdot 5, 5 \cdot 7, 5 \cdot 23, 7 \cdot 23)$, but no divisors using one of the prime factors at a time. (Why?) Hence, $k = 8$.

Challenge 1   *Find the largest integer that divides 364, 414, and 539 with the same remainder in each case.*

Let D be the largest integer dividing the given numbers with the same remainder. Then $364 = D \times Q_1 + R$, $414 = D \times Q_2 + R$, $539 = D \times Q_3 + R$. $\therefore 50 = D(Q_2 - Q_1)$, $125 = D(Q_3 - Q_2)$, $175 = D(Q_3 - Q_1)$. Since D is an exact (largest) divisor of 50, 125, and 175, $D = 25$.

Check to see that the remainders are equal.

Challenge 2 *A somewhat harder problem is this: find the largest integer that divides 364, 414, and 541 with remainders R_1, R_2, and R_3, respectively, such that $R_2 = R_1 + 1$, and $R_3 = R_2 + 1$.*

Using a similar procedure, we have $364 = DQ_1 + R_1$, $414 = DQ_2 + R_1 + 1$, $541 = DQ_3 + R_1 + 1 + 1$. $\therefore 50 = D(Q_2 - Q_1) + 1$, $127 = D(Q_3 - Q_2) + 1$, $177 = D(Q_3 - Q_1) + 2$. $\therefore 49 = D(Q_2 - Q_1)$, $126 = D(Q_3 - Q_2)$, $175 = D(Q_3 - Q_1)$. $\therefore D = 7$

Challenge 3 *A committee of three students, A, B, and C, meets and agrees that A report back every 10 days, B, every 12 days, and C, every 15 days. Find the least number of days before C again meets both A and B.*

We must find the smallest integer exactly divisible by 10, 12, and 15; that is, the least common multiple M of 10, 12, and 15.

Since $10 = 2 \cdot 5$, $12 = 2 \cdot 2 \cdot 3$, and $15 = 3 \cdot 5$, $M = 2 \cdot 5 \cdot 2 \cdot 3 = 60$. That is, A, B, and C will be together again in 60 days. Put another way, C will find both A and B on the fourth time that he reports back.

2-11 *List all the possible remainders when an even integer square is divided by 8.*

First we observe that if N^2 is even, then N is even. (Justify this statement.) Therefore, $N = 2k$ where k is an even or an odd integer.

When k is odd, say $2m + 1$, then $N^2 = 4k^2 = 4(4m^2 + 4m + 1) = 16m(m + 1) + 4$, and the remainder, upon division by 8, is 4.

When k is even, say $2m$, then $N^2 = 4k^2 = 4(4m^2) = 16m^2$, and the remainder, upon division by 8, is 0. Therefore, the possible remainders are either 0 or 4.

2-12 *Which is larger: the number of partitions of the integer* $N = k \cdot 10^2$ *into* $2k + 1$ *positive even integers, or the number of partitions of* N *into* $2k + 1$ *positive odd integers, where* $k = 1, 2, 3, \ldots$? *To partition a positive integer is to represent the integer as a sum of positive integers.*

Since $N = k \cdot 10^2$ is even, it cannot be partitioned into $2k + 1$ positive odd integers since $2k + 1$ is odd. Therefore, the partitioning into even integers is larger.

2-13 *Given the three-digit number* $N = a_1a_2a_3$, *written in base* 10, *find the least absolute values of* m_1, m_2, m_3 *such that* N *is divisible by 7 if* $m_1a_1 + m_2a_2 + m_3a_3$ *is divisible by 7.*

$$N = a_1 \cdot 10^2 + a_2 \cdot 10 + a_3 = a_1(7 + 3)^2 + a_2(7 + 3) + a_3$$
$$= 7(14a_1 + a_2) + 2a_1 + 3a_2 + a_3.$$

When N is divided by 7, the remainder is $2a_1 + 3a_2 + a_3$. It follows that N is exactly divisible by 7 if $2a_1 + 3a_2 + a_3$ is exactly divisible by 7.

Therefore, $m_1 = 2, m_2 = 3, m_3 = 1$.

Challenge 1 *Solve the problem for the six-digit number*

$$N = a_1a_2a_3a_4a_5a_6.$$

NOTE: Only $|m_1|$, $|m_2|$, and $|m_3|$ are needed.

$$N = a_6 + a_5(7 + 3) + a_4(7 + 3)^2 + a_3(7 + 3)^3$$
$$+ a_2(7 + 3)^4 + a_1(7 + 3)^5$$

Therefore,

$$N = 7p_1 + a_6 + 3a_5 + 9a_4 + 27a_3 + 81a_2 + 243a_1$$
$$= 7p_2 + a_6 + 3a_5 + 2a_4 + 6a_3 + 4a_2 + 5a_1$$
$$= 7p_3 + a_6 + 3a_5 + 2a_4 - a_3 - 3a_2 - 2a_1.$$

Therefore, N is exactly divisible by 7 if

$$-2a_1 - 3a_2 - a_3 + 2a_4 + 3a_5 + a_6$$

is exactly divisible by 7. Hence, the least absolute values of m_1, m_2, m_3 are 2, 3, and 1 respectively.

2-14 *When* $x^3 + a$ *is divided by* $x + 2$, *the remainder is known to be* -15. *Find the numerical value of* a.

METHOD I: Using the Remainder Theorem (see Appendix II), we have $(-2)^3 + a = -15$, $a = -7$.

METHOD II: By long division, $(x^3 + a) \div (x + 2)$ yields the remainder $a - 8$. $\therefore a - 8 = -15$, $a = -7$

2-15 *If* $x - a$ *is a factor of* $x^2 + 2ax - 3$, *find the numerical value(s) of* a.

METHOD I: Let the second factor be $x + b$. Then

$$x^2 + 2ax - 3 = (x - a)(x + b) = x^2 + x(-a + b) - ab.$$

$\therefore 3 = ab$, and $2a = -a + b$, $\therefore a = +1$ or -1, and $b = +3$ or -3, respectively. Check $x^2 + 2x - 3 = (x - 1)(x + 3)$, and $x^2 - 2x - 3 = (x + 1)(x - 3)$.

METHOD II: With the use of the Remainder Theorem and the Factor Theorem (see Appendix II), we have $P(x) = x^2 + 2ax - 3 = 0$ when $x = a$. Since $x - a$ is a factor, $a^2 + 2a^2 - 3 = 0$, $3a^2 = 3$, $a = +1$ or -1.

Challenge 1 *Find the remainder when* $P(x) = x^3 - 2x^2 + 2x - 2$ *is divided by* $x + 1$.

One way to find out is to perform a long division to obtain the remainder. The quicker way is to use the Remainder Theorem (see Appendix II):

$$P(-1) = (-1)^3 - 2(-1)^2 + 2(-1) - 2 = -7.$$

2-16 *Let* N *be the product of five different odd prime numbers. If* N *is the five-digit number* abcab, $4 < a < 8$, *find* N.

To obtain some idea of the size of the primes involved, note that since 11^5 is a six-digit number, some of the primes are less than 11, and at least one is more than 11. We could guess at the first five odd primes, but $3 \cdot 5 \cdot 7 \cdot 11 \cdot 13 = 15{,}015$, which is unacceptable since it is required that a be greater than 4. Our second trial could very well be $5 \cdot 7 \cdot 11 \cdot 13 \cdot 17 = 85{,}085$, which is unacceptable since it is required that $a < 8$.

Close examination reveals that *abcab* must be a multiple of $100 = 7 \cdot 11 \cdot 13$. Since the only numbers between 51 and 79 which have exactly two odd prime factors different from 7, 11, and 13 are 51, 57, and 69, the values of N are: $51051 = 3 \cdot 7 \cdot 11 \cdot 13 \cdot 17$; $57057 = 3 \cdot 7 \cdot 11 \cdot 13 \cdot 19$; and $69069 = 3 \cdot 7 \cdot 11 \cdot 13 \cdot 23$.

2-17 *If a five-digit number* N *is such that the sum of the digits is* 29, *can* N *be the square of an integer?*

Assume that $N = (h \cdot 10^2 + t \cdot 10 + u)^2$. Since the remainder obtained when dividing an integer by 9 is equal to the remainder obtained when dividing the sum of its digits by 9 (reduced by multiples of 9, if necessary), the remainder for N is equal to the remainder for $(h + t + u)^2$, mod 9. However, while the given sum 29 yields a remainder of 2 when divided by 9, the only remainders possible with squares of integers are 0, 1, 4, and 7. (To verify this last remark, designate all integers as n, $n + 1$, $n + 2, \ldots n + 8$, with n divisible by 9. Square these expressions and examine the result.) Consequently, N cannot be the square of an integer.

ILLUSTRATION 1: $N = 24{,}689$ is not the square of an integer; the sum of its digits is 29.

ILLUSTRATION 2: $N = 24{,}649 = 157^2$; the sum of its digits is 25, which, divided by 9, yields a remainder of 7.

Note that the converse of this theorem is not true.

ILLUSTRATION 3: $N = 24{,}694$ is not the square of an integer. Yet the sum of the digits is 25 with a remainder of 7 when divided by 9.

2-18 *Each of the digits* 2, 3, 4, 5 *is used once and once only in writing a four-digit number. Find the number of such numbers and their sum.*

For the thousands' position, we may choose any one of the given four digits; for the hundreds' position, any one of the remaining three; for the tens' position, either one of the remaining two. The units' digit is assigned the fourth of the given digits. In all, there are $4 \times 3 \times 2 \times 1 = 24$ possibilities.

Each digit, then, appears 6 times in each position. Therefore, the sum is

$$6(5555) + 6(4444) + 6(3333) + 6(2222) = 6(15{,}554) = 93{,}324.$$

2-19 *Find all positive integral values of* k *for which* 8k + 1 *expressed in base* 10 *exactly divides* 231 *expressed in base* 8.

$231_8 = 2 \cdot 8^2 + 3 \cdot 8 + 1 = 153_{10} = 9 \cdot 17 = 1 \cdot 153$. Notice that $9 = 1 \cdot 8 + 1$, $17 = 2 \cdot 8 + 1$, $1 = 0 \cdot 8 + 1$, and $153 = 19 \cdot 8 + 1$. Restricted to positive values, $k = 1, 2, 19$.

2-20 *Express in terms of* n *the positive geometric mean of the positive divisors of the natural number* n. *Definition: the positive geometric mean of the* k *positive numbers* $a_1, a_2, \ldots, a_k$ *is* $\sqrt[k]{a_1 a_2 \ldots a_k}$.

Let the divisors of n, arranged in increasing order, be $d_1, d_2, \ldots, d_r$. Then $n = d_1 d_r = d_2 d_{r-1} = d_3 d_{r-2} = \cdots = d_i d_{r-i+1}$.
$d_1 d_r d_2 d_{r-1} \ldots d_r d_1 = d_1{}^2 d_2{}^2 \ldots d_r{}^2 = n^r$
$\therefore \sqrt[r]{d_1{}^2 d_2{}^2 \ldots d_r{}^2} = \sqrt[r]{n^r} = n$, and $\sqrt[r]{d_1 d_2 \ldots d_r} = \sqrt{n}$.

ILLUSTRATION 1: The divisors of 16 are 1, 2, 4, 8, 16, five in all.
$$\sqrt[5]{1 \cdot 2 \cdot 4 \cdot 8 \cdot 16} = \sqrt[5]{4^5} = 4 = \sqrt{16}$$

ILLUSTRATION 2: The divisors of 24 are 1, 2, 3, 4, 6, 8, 12, 24, eight in all.
$$\sqrt[8]{1 \cdot 2 \cdot 3 \cdot 4 \cdot 6 \cdot 8 \cdot 12 \cdot 24 \cdot} = \sqrt[8]{24^4} = \sqrt{24}$$

---

3 Relations: Familiar and Surprising

3-1 *Let* $y_1 = \dfrac{x+1}{x-1}$. *Let* y_2 *be the simplified expression obtained by replacing* x *in* y_1 *by* $\dfrac{x+1}{x-1}$. *Let* y_3 *be the simplified expression obtained by replacing* x *in* y_2 *by* $\dfrac{x+1}{x-1}$, *and so forth. Find* y_6, y_{100}, y_{501}.

This looks frighteningly difficult, but it isn't!

$$y_2 = \frac{\dfrac{x+1}{x-1} + 1}{\dfrac{x+1}{x-1} - 1} = \frac{x+1+x-1}{x+1-x+1} = \frac{2x}{2} = x$$

$$\therefore y_3 = \frac{x+1}{x-1}, \quad y_4 = x,$$

and so forth. For all even subscripts, the value of $y_{2k} = x$, for all odd subscripts the value $y_{2k+1} = \dfrac{x+1}{x-1} \therefore y_6 = x, y_{100} = x$, $y_{501} = \dfrac{x+1}{x-1}$.

3-2 *Let us designate a* lattice point *in the rectangular Cartesian plane as one with integral coordinates. Consider a rectangle with sides parallel to the axes such that there are* s_1 *lattice points in the base and* s_2 *lattice points in the altitude, and that the vertices are lattice points.*

(a) *Find the number of interior lattice points,* $N(I)$.
(b) *Find the number of boundary lattice points,* $N(B)$.
(c) *Find the total number of lattice points,* N.

(a) $N(I) = (s_1 - 2)(s_2 - 2) = s_1 s_2 - 2s_1 - 2s_2 + 4$
(b) $N(B) = 2s_1 + 2(s_2 - 2) = 2s_1 + 2s_2 - 4$. An alternative form for $N(B)$ is $2[(s_1 - 1) + (s_2 - 1)]$.
(c) $N = N(I) + N(B) = s_1 s_2$

3-3 *An approximate formula for a barometric reading,* p(*millimeters*), *for altitudes* h(*meters*) *above sea level, is* $p = 760 - .09h$, *where* $h \leq 500$. *Find the change in* p *corresponding to a change in* h *from* 100 *to* 250.

Since $p_1 = 760 - .09h_1$ and $p_2 = 760 - .09h_2$, then $p_1 - p_2 = -.09(h_1 - h_2)$, or $\Delta p = -.09 \Delta h$ where Δp is the change in p and Δh is the change in h. $\therefore \Delta p = -.09(150) = -13.5$ mm.; that is, a decrease of 13.5 mm. in barometric pressure.
Check by finding p_1 for $h = 100$, and p_2 for $h = 250$.

3-4 *A student wishing to give 25 cents to each of several charities finds that he is 10 cents short. If, instead, he gives 20 cents to each of the charities, then he is left with 25 cents. Find the amount of money with which the student starts.*

Let n represent the number of charities, and A, in cents, the amount of money with which the student starts.
The first condition, translated, becomes $A - 25n = -10$.
The second condition, translated, becomes $A - 20n = 25$.
Therefore, $5n = 35$, $n = 7$. $A - 20 \cdot 7 = 25$, $A = 165$. The amount started with is $1.65.

Challenge 3 *How does the answer change if the original shortage is 25 cents?*

In a certain sense this is an unfair question. He could simply reduce the number of charities (if unspecified) by one. In this case, the amount A is undetermined, except to say that it is a multiple of 25. If, however, the number of charities is fixed, then the amount A is $2.25.

3-5 *Find two numbers* x *and* y *such that* xy, $\frac{x}{y}$, *and* x − y *are equal.*

$xy = \frac{x}{y}$ ∴ $y = 1$ or -1. Also $x - y = xy$. For $y = 1$, $x - 1 = x$, a contradiction. For $y = -1$, $x + 1 = -x$, $x = -\frac{1}{2}$, ∴ $x = -\frac{1}{2}, y = -1$.

3-6 *A merchant on his way to the market with* n *bags of flour passes through three tollgates. At the first gate, the toll is* $\frac{1}{4}$ *of his holdings, but 3 bags are returned. At the second gate, the toll is* $\frac{1}{3}$ *of his (new) holdings, but 2 bags are returned. At the third gate, the toll is* $\frac{1}{2}$ *of his (new) holdings, but 1 bag is returned. The merchant arrives at the market with exactly* $\frac{n}{2}$ *bags. If all transactions involve whole bags, find the value of* n.

The number of bags remaining after the first toll is $\frac{3n}{4} + 3$, after the second toll, $\frac{n}{2} + 4$, and after the third toll, $\frac{n}{4} + 3$. ∴ $\frac{n}{4} + 3 = \frac{n}{2}$, $n = 12$.

3-7 *The number* N_2 *is* 25% *more than the number* N_1, *the number* N_3 *is* 20% *more than* N_2, *and the number* N_4 *is* x% *less than* N_3. *For what value of* x *is* $N_4 = N_1$?

$N_4 = N_3 \left(1 - \frac{x}{100}\right) = \left(1 - \frac{x}{100}\right)\left(1 + \frac{20}{100}\right) N_2$

$= \left(1 - \frac{x}{100}\right)\left(1 + \frac{20}{100}\right)\left(1 + \frac{25}{100}\right) N_1$

For N_4 to equal N_1, $\left(1 - \frac{x}{100}\right)\left(1 + \frac{1}{5}\right)\left(1 + \frac{1}{4}\right)$ must equal 1.

There, $\left(1 - \frac{x}{100}\right)\left(\frac{6}{5}\right)\left(\frac{5}{4}\right) = 1$; $1 - \frac{x}{100} = \frac{2}{3}$, $x = 33\frac{1}{3}$.

3-8 *Let* R = px *represent the revenue,* R *(dollars), obtained from the sale of* x *articles, each at selling price* p *(dollars). Let* C = mx + b *represent the total cost,* C, *in dollars, of producing and selling these* x *articles. How many articles must be sold to break even?*

At the break-even point, $R = C$; that is, $px = mx + b$, so that $x = \frac{b}{p - m}$. For this value of x, $R = C = \frac{pb}{p - m}$.

3-9 *In a certain examination it is noted that the average mark of those passing is 65, while the average mark of those failing is 35. If the average mark of all participants is 53, what percentage of the participants passed?*

We represent the number of participants passing by P, and the number failing by F. The total score of those passing is $65P$, the total score of those failing is $35F$, and the total score of all is $53(P + F)$.

$$65P + 35F = 53(P + F), \quad 12P = 18F, \quad F = \frac{2}{3}P.$$ Since the ratio of those passing to all the participants is $\frac{P}{P + F}$, we have $\frac{P}{P + F} = \frac{P}{P + \frac{2}{3}P} = \frac{3}{5}$. The percentage passing is $\frac{3}{5}(100) = 60$.

3-10 *Under plan I, a merchant sells n_1 articles, priced 1 for 2¢, with a profit of $\frac{1}{4}$¢ on each article, and n_2 articles, priced 2 for 3¢, with a profit of $\frac{1}{8}$¢ on each article. Under Plan II, he mixes the articles and sells them at 3 for 5¢. If $n_1 + n_2$ articles are sold under each plan, for what ratio $\frac{n_1}{n_2}$ is the profit the same?*

Under Plan I, the selling price is $2n_1 + \frac{3}{2}n_2$, the profit is $\frac{1}{4}n_1 + \frac{1}{8}n_2$, and the cost is $1\frac{3}{4}n_1 + 1\frac{3}{8}n_2$. Under Plan II, the selling price is $\frac{5}{3}(n_1 + n_2)$, the cost is the same as in Plan I, and the profit is $-\frac{1}{12}n_1 + \frac{7}{24}n_2$.

Therefore, $\frac{n_1}{4} + \frac{n_2}{8} = -\frac{n_1}{12} + \frac{7n_2}{24}$, $\frac{n_1}{3} = \frac{n_2}{6}$, $\frac{n_1}{n_2} = \frac{1}{2}$. We could just as well have set the Plan I selling price equal to the Plan II selling price to obtain the required ratio since, if the profit is the same and the number of each article sold is the same, the selling price must be the same under both plans. Work out the details.

Challenge *Change 2¢ to p¢ and 3¢ to q¢ and solve the problem.*

Under Plan I, the selling price is $pn_1 + \frac{qn_2}{2}$, and under Plan II, the selling price is $\frac{p + q}{3}(n_1 + n_2)$. Since the costs are the same, and the profits are the same under both plans, it follows that $pn_1 + \frac{qn_2}{2} = \frac{p+q}{3}(n_1 + n_2)$.

$$\therefore n_1\left(\frac{2p-q}{3}\right) = n_2\left(\frac{2p-q}{6}\right), \frac{n_1}{n_2} = \frac{1}{2}$$

3-11 *The sum of two numbers* x *and* y, *with* x > y, *is 36. When* x *is divided by 4 and* y *is divided by 5, the sum of the quotients is 8. Find the numbers* x *and* y.

Working formally, we have $x + y = 36$, $\frac{x}{4} + \frac{y}{5} = 8$. From this pair of equations we obtain $x = 16$, $y = 20$. But it is given that $x > y$. Consequently, there is no solution to the problem.

3-12 *Find the values of* x *satisfying the equation* $|x - a| = |x - b|$, *where* a, b *are distinct real numbers.*

The interpretation $x - a = x - b$ contradicts the given information that $a \neq b$.

Hence, $x - a = -(x - b)$, or $-(x - a) = x - b$. In either case, $2x = a + b$, $x = \frac{1}{2}(a + b)$; that is, x is the arithmetic mean between a and b.

Here, again, a geometric interpretation is enlightening. If $a < b$, it follows that $a < x < b$; point a is to the left of point x which is to the left of point b. The distance $x - a$ equals the distance $b - x$. If $a > b$, it follows that $b < x < a$; the order of points from left to right is b, x, a. Again the distance $x - b$ equals the distance $a - x$. A single expression for both cases is $|x - a| = |x - b|$.

Challenge 3 *Find the values of* x *satisfying the equation* $|2x - 1| = |x - 2|$.

Here a new element requires our consideration. If we think of $|2x - 1|$ as $2\left|x - \frac{1}{2}\right|$, we interpret the problem to mean that the distance of x from 2 is twice the distance of x from $\frac{1}{2}$. This allows for the two possibilities: $2\left(x - \frac{1}{2}\right) = x - 2$, and $2\left(x - \frac{1}{2}\right) = -(x - 2)$. In the former case $x = -1$, in the latter, $x = 1$. (See Figs. S3-12a, and S3-12b.)

S3-12a   S3-12b

3-13 *Two night watchmen, Smith and Jones, arrange for an evening together away from work. Smith is off duty every eighth evening starting today, while Jones is off duty every sixth evening starting tomorrow. In how many days from today can they get together?*

Smith is off the first day, the ninth day, the seventeenth day, and so forth; that is, on the days numbered $1 + 8S$, where $S = 0, 1, 2, \ldots$. Similarly, Jones' days off may be represented by $2 + 6J$, where $J = 0, 1, 2, \ldots$.

We must, therefore, find a solution in positive integers to the equation $1 + 8S = 2 + 6J$, or $8S = 1 + 6J$. Since, for integer values of S and J, the left side is always even while the right side is always odd, the equation is not solvable in integers.

Smith and Jones cannot get together.

3-14 *A man buys 3-cent stamps and 6-cent stamps, 120 in all. He pays for them with a $5.00 bill and receives 75 cents in change. Does he receive the correct change?*

Represent by x the number of 3-cent stamps. Then $120 - x$ represents the number of 6-cent stamps. The total cost of the stamps, in cents, is $C = 3x + 6(120 - x) = 720 - 3x$, so that C is divisible by 3. Hence, the payment P should be divisible by 3. But $P = 500 - 75 = 425$ is not divisible by 3. It follows that the 75 cents change is incorrect.

3-15 *In how many ways can a quarter be changed into dimes, nickels, and cents?*

As the problem is stated it is somewhat ambiguous. We want to know if we must use three coin-types, or if we are permitted to use two coin-types, or one coin-type. We consider each case.

Representing the number of dimes, nickels, and cents respectively by d, n, and c, we have $25 = 10d + 5n + c$.

The equation contains three unknown quantities, but we have only this one equation. Is there any other helpful information?

Yes, the knowledge that each of d, n, and c is a positive integer (or zero for the second and third cases).

This is an instance of what mathematicians call a Diophantine equation. To a large extent it is solved by trial. Obviously, if $d = 0$, $n = 0$, then $c = 25$. The table below shows all possible combinations.

d	0	0	0	0	0	0	1	1	1	1	2	2
n	0	1	2	3	4	5	0	1	2	3	0	1
c	25	20	15	10	5	0	15	10	5	0	5	0

Satisfy yourself that no permissible combination has been omitted. Therefore, if at least one of each coin must be used, there are just two possibilities. If only two coin-types are used, there are eight possibilities. If a single coin-type is acceptable, there are, again, just two possibilities. The total for the three cases is 12 possibilities.

Challenge   *Is the answer unique if it is stipulated that there are five times as many coins of one kind as of the other two kinds together?*

Of course if a list of all possibilities (as shown above) is available, we merely read off the answer. How do we proceed if no such list is available, and we do not care to prepare one?

We immediately rule out the possibility of 5 dimes or 5 nickels. That leaves only the possibility that the number of cents is five times the combined number of dimes and nickels.

$$\therefore 25 = 10d + 5n + 5(d + n),$$
$$25 = 15d + 10n, \quad 5 = 3d + 2n.$$

Obviously, $d = 1$, $n = 1$. The combination of 1 dime, 1 nickel, 10 cents is unique.

3-16 *Find the number of ways in which 20 U.S. coins, consisting of quarters, dimes, and nickels, can have a value of $3.10.*

Letting q, d, and n, respectively, represent the number of quarters, dimes, and nickels, we translate one condition of the problem into $25q + 10d + 5n = 310$. Since there are 20 coins in all, $n = 20 - q - d$.

Substituting for n in the first equation and simplifying, we have the relatively simple equation $4q + d = 42$. The equation $4q + d = 42$ could be solved by trial and error, but it will probably save time to proceed as follows.

Solving for q we have $q = \dfrac{42 - d}{4} = 10 - \dfrac{d - 2}{4}$. To insure an integral value for q, the quantity $d - 2$ must be a multiple of 4. Set $d - 2 = 4k$. Then $q = 10 - k$ where $k = 0, 1, 2, \ldots$. Taking, in turn, $k = 0, 1, 2, \ldots$ we have the following.

k	d	q	n	total
0	2	10	8	20
1	6	9	5	20
2	10	8	2	20
3	14	7	−1	(Reject)

Note the result of taking $k \geq 3$. There are, therefore, three acceptable ways, as shown in rows 1, 2, 3.

An alternate method for solving the Diophantine equation $4q + d = 42$ is as follows. Since the greatest common factor of 4 and 1 exactly divides 42 there exist integer solutions to the equation.

By observation, one such solution is $q = 10$, $d = 2$. As a result of subtracting $4(10) + (2) = 42$ from $4q + d = 42$, we get $4(q - 10) + d - 2 = 0$. Hence $\dfrac{q - 10}{1} = \dfrac{-d + 2}{4} = t$, where t is an integer. Therefore, $q = t + 10$, and $d = 2 - 4t$. Taking integral values of t we have the following.

t	d	q	n	total
1	−2	11	0	(Reject)
0	2	10	8	20
−1	6	9	5	20
−2	10	8	2	20
−3	14	7	−1	(Reject)

Note the result of taking $t \geq 1$, or $t \leq -3$ yields an absurd answer; thus there are only three possible answers.

4 Bases: Binary and Beyond

4-1 *Can you explain mathematically the basis for the following correct method of multiplying two numbers, sometimes referred to as the Russian Peasant Method of multiplication?*

Let us say that we are to find the product 19 × 23. *In successive rows, we halve the entries in the first column, rejecting the remainders of* 1 *where they occur. In the second column, we double each successive entry. This process continues until a* 1 *appears in column I.*

I	II
19	23
9	46
4	92
2	184
1	368
	437

We then add the entries in column II, omitting those that are associated with the even entries in column I.

$$(19)(23) = (9 \cdot 2 + 1)(23) = 9 \cdot 46 + 1 \cdot 23$$
$$(9)(46) = (4 \cdot 2 + 1)(46) = 4 \cdot 92 + 1 \cdot 46$$
$$(4)(92) = (2 \cdot 2 + 0)(92) = 2 \cdot 184 + 0 \cdot 92$$
$$(2)(184) = (1 \cdot 2 + 0)(184) = 1 \cdot 368 + 0 \cdot 184$$
$$(1)(368) = (0 \cdot 2 + 1)(368) = \underline{1 \cdot 368}$$
$$437$$

The binary nature of this multiplication is shown in the following.

$$(19)(23) = (1 \cdot 2^4 + 0 \cdot 2^3 + 0 \cdot 2^2 + 1 \cdot 2 + 1)(23)$$
$$= 1 \cdot 23 + 2 \cdot 23 + 0 \cdot 23 + 0 \cdot 23 + 2^4 \cdot 23$$
$$= 23 + 46 + 0 + 0 + 368 = 437$$

4-2 *If* x = {0, 1, 2, . . . , n, . . .}, *find the possible terminating digits of* $x^2 + x$ *in base 2.*

Whether x is odd or even, $x^2 + x = x(x + 1)$ is even since, if x is odd, $x + 1$ is even, and if $x + 1$ is odd, x is even. Therefore, in every instance, the terminating digit is 0.

4-3 *Find the base* b *such that* $72_b = 2(27_b)$. 72_b *means* 72 *written in base* b.

$7b + 2 = 2(2b + 7) = 4b + 14$ ∴ $3b = 12$, $b = 4$. This value is unacceptable since there is no digit 7 in base 4.

Challenge 1   *Try the problem for* $73_b = 2(37_b)$.

$7b + 3 = 2(3b + 7)$ ∴ $b = 11$

4-4 *In what base* b *is* 441_b *the square of an integer?*

$441_b = 4b^2 + 4b + 1 = (2b + 1)^2$. Therefore, 441_b is the square of an integer in all bases $b > 4$, as 4 must be a member of the set of digits to be used in base b.

ILLUSTRATION 1: $441_5 = (21_5)^2$

ILLUSTRATION 2: $441_{10} = (21_{10})^2$

ILLUSTRATION 3: $441_{12} = (21_{12})^2$

Challenge 1   *If* N *is the base* 4 *equivalent of* 441 *written in base* 10, *find the square root of* N *in base* 4.

$$441_{10} = 12321_4 = 1 \cdot 4^4 + 2 \cdot 4^3 + 3 \cdot 4^2 + 2 \cdot 4 + 1$$
$$= (1 \cdot 4^2 + 1 \cdot 4 + 1)^2 = N$$
$$\therefore \sqrt{N} = 111_4$$

Challenge 2   *Find the smallest base* b *for which* 294_b *is the square of an integer.*

$294_b = 2b^2 + 9b + 4 = (2b + 1)(b + 4)$

Since 294_b is even and $2b + 1$ is odd, then $b + 4$ is even so that b is even, and $b \geq 10$. (Why?) It follows that each factor is the square of an integer.

If $b = 10$, then $2b + 1 = 21$, not the square of an integer.

If $b = 12$, then $2b + 1 = 25 = 5^2$ and $b + 4 = 16 = 4^2$.

VERIFICATION: $294_{12} = 2 \cdot 12^2 + 9 \cdot 12 + 4 = (1 \cdot 12 + 8)^2 = (18_{12})^2$

COMMENT: The next larger base is 60. Verify.

4-5 *Let* N *be the three-digit number* $a_1a_2a_3$ *written in base* b, b $\geq$ 2, *and let* S $=$ $a_1 + a_2 + a_3$. *Prove that* N $-$ S *is divisible by* b $-$ 1.

PROOF:

$$N = a_1 \cdot b^2 + a_2 \cdot b + a_3$$
$$= a_1(b - 1 + 1)^2 + a_2(b - 1 + 1) + a_3$$
$$= a_1(b - 1)^2 + 2a_1(b - 1) + a_1 + a_2(b - 1) + a_2 + a_3$$
$$\therefore N - S = a_1(b - 1)^2 + 2a_1(b - 1) + a_2(b - 1)$$

Since the right side of this last equality is divisible by $b - 1$, so is the left side.

4-6 *Let* N *be the four-digit number* $a_0a_1a_2a_3$ *(in base* 10*), and let* N′ *be the four-digit number which is any of the* 24 *rearrangements of the digits. Let* D $=$ |N $-$ N′|. *Find the largest digit that exactly divides* D.

Since $N = a_0 \cdot 10^3 + a_1 \cdot 10^2 + a_2 \cdot 10 + a_3 = a_0(9 + 1)^3 + a_1(9 + 1)^2 + a_2(9 + 1) + a_3$, we can express N as $9K + a_0 + a_1 + a_2 + a_3$. Similarly, $N' = 9K' + a_0 + a_1 + a_2 + a_3$. Therefore, $D = |9K - 9K'| = 9|K - K'|$, so that D is exactly divisible by 9.

4-7 *Express in binary notation (base* 2*) the decimal number* 6.75.

This one is easy enough to do without a formal procedure. The fractional part $.75 = \frac{3}{4}$ and can be expressed as $\frac{1}{2} + \frac{1}{2^2}$. The integral part 6 is equivalent to $1 \cdot 2^2 + 1 \cdot 2 + 0$.

Therefore, $6.75 = 1 \cdot 2^2 + 1 \cdot 2 + 0 + \frac{1}{2} + \frac{1}{2^2}$
$$= 1 \cdot 2^2 + 1 \cdot 2 + 0 \cdot 2^0 + 1 \cdot 2^{-1} + 1 \cdot 2^{-2},$$
so that 6.75 (base 10) = 110.11 (base 2).

Challenge 1 *Convert the decimal number* N $=$ 19.65625 *into a binary number.*

For less simple cases we may need a formal procedure.
(a) For the integral part of N, 19, we obtain the non-negative integral powers of two as follows.

$$19 = a_k + 2(a_{k-1}) + 2^2(a_{k-2}) + \cdots + 2^{k-1}(a_1) \quad \text{(I)}$$

Divide equation (I) by 2; the remainder 1 of the left side equals the remainder a_k of the right side. Removing these remainders we now have

$$9 = a_{k-1} + 2(a_{k-2}) + 2^2(a_{k-3}) + \cdots. \quad \text{(II)}$$

Divide equation (II) by 2; again the remainder 1 of the left side equals the remainder a_{k-1} of the right side. Removing these remainders we now have

$$4 = a_{k-2} + 2(a_{k-3}) + 2^2(a_{k-4}) + \cdots. \quad \text{(III)}$$

Divide equation (III) by 2; the left side remainder 0 equals the right side remainder a_{k-2}. Continuing in this manner we find that the process ends in two more steps with $a_{k-3} = 0$ and $a_{k-4} = 1$.

Reassembling these partial results we have

$$19 = 1 \cdot 2^4 + 0 \cdot 2^3 + 0 \cdot 2^2 + 1 \cdot 2 + 1$$
$$= 10011 \ (\text{base } 2).$$

(b) For the fractional part of N, .65625, we obtain the negative integral powers of 2 as follows. Let $0.65625 = \frac{b_1}{2} + \frac{b_2}{2^2} + \frac{b_3}{2^3} + \cdots$. For convenience, however, we reduce $\frac{65625}{100000}$ to $\frac{21}{32}$ so that

$$\frac{21}{32} = \frac{b_1}{2} + \frac{b_2}{2^2} + \frac{b_3}{2^3} + \cdots.$$

Multiplying the equation above by 2, we obtain $\frac{42}{32} = 1 + \frac{10}{32} = 1 + \frac{5}{16}$, so that $b_1 = 1$. Multiplying $\frac{5}{16}$ by 2, we have $\frac{10}{16} = 0 + \frac{5}{8}$, so that $b_2 = 0$. Multiplying $\frac{5}{8}$ by 2, we have $\frac{10}{8} = 1 + \frac{1}{4}$, so that $b_3 = 1$. Multiplying $\frac{1}{4}$ by 2, we have $0 + \frac{1}{2}$, so that $b_4 = 0$. Finally, $\frac{1}{2} \times 2 = 1$, so that $b_5 = 1$, and the process ends.

Reassembling these partial results we find $0.65625 = \frac{1}{2} + \frac{0}{2^2} + \frac{1}{2^3} + \frac{0}{2^4} + \frac{1}{2^5} = .10101 \ (\text{base } 2).$

Therefore, 19.65625 (base 10) = 10011.10101 (base 2).

Challenge 2 *Does the (base 10) non-terminating expansion 5.333 . . .
terminate when converted into base 2?*

As a first consideration we ask ourselves, which
reduced proper fractions have decimal expansions that
terminate? Answer: Those whose denominators contain
only the factors 2 and 5, exact divisors of 10. It would
seem reasonable to conclude that reduced fractions whose
denominators contain only the factors 2 will have
terminating expansions in the base 2.

This was illustrated above with the decimal 0.65625,
which became .10101 in the base 2. Additional illustrations follow.

ILLUSTRATION 1: We return to the problem above and
convert the decimal 5.333 . . . into a "binimal" [the word
binimal is an *ad hoc* invention].

$$5 = 1 \cdot 2^2 + 0 \cdot 2 + 1 \therefore 5 \text{ (base 10)} = 101 \text{ (base 2)}$$
$$.333 \ldots = \frac{1}{3}, \frac{1}{3} \times 2 = 0 + \frac{2}{3}, \frac{2}{3} \times 2 = 1 + \frac{1}{3}$$

From this point on the digits 0 and 1 repeat endlessly.
Therefore, 5.333 . . . (base 10) = 101.010101 . . . (base 2).
This establishes that 5.333 . . . (base 10) does not termi-
nate when converted to base 2.

ILLUSTRATION 2: Even a terminating expansion in base 10
may become non-terminating in base 2. For example,
8.60 (base 10) = 1000.10011001 . . . (base 2).

ILLUSTRATION 3: Convert the decimal 8.60 into a quinimal
(base 5).
ANSWER: 8.60 (base 10) = 13.3 (base 5)

ILLUSTRATION 4: Convert the decimal 8.60 into a senimal
(base 6).
ANSWER: 8.60 (base 10) = 12.333 . . . (base 6)

ILLUSTRATION 5: Convert the decimal 5.333 . . . into a
senimal.
ANSWER: 5.2

ILLUSTRATION 6: Convert the binimal 111.001 into a
senimal.
ANSWER: 11.043

4-8 *Assume* $r = \{6, 7, 8, 9, 10\}$ *and* $1 < a < r$. *If there is exactly* one *integer value of* a *for which* $\frac{1}{a}$, *expressed in the base* r, *is a terminating* r-mal, *find* r.

Try, in turn, each member of the set r. For $r = 6$, there are 2 terminating r-mals, to wit, $\frac{1}{2} = \frac{3}{6}$ and $\frac{1}{3} = \frac{2}{6}$.

For $r = 7$, there are no terminating r-mals.

For $r = 8$, there are 2 terminating r-mals, to wit, $\frac{1}{2} = \frac{4}{8}$ and $\frac{1}{4} = \frac{2}{8}$.

For $r = 9$, there is just 1 terminating r-mal, to wit, $\frac{1}{3} = \frac{3}{9}$.

For $r = 10$, there are 2 terminating r-mals, to wit, $\frac{1}{2} = \frac{5}{10}$ and $\frac{1}{5} = \frac{2}{10}$.

The answer is, therefore, $r = 9$.

4-9 *From the unit segment* $\overline{OA}$ *extending from the origin* O *to* A(1, 0), *remove the middle third. Label the remaining segments* $\overline{OB}$ *and* $\overline{CA}$, *and remove the middle third from segment* $\overline{OB}$. *Label the first two remaining segments* $\overline{OD}$ *and* $\overline{EB}$. *Express the coordinates of* D, E, *and* B *in base* 3.

$OB = \frac{1}{3}$, so that the coordinates of B in base 3 are (.1, 0).

$OD = \frac{1}{9} = \frac{0}{3} + \frac{1}{3^2}$, so that the coordinates of D in base 3 are (.01, 0).

$OE = \frac{2}{9} = \frac{0}{3} + \frac{2}{3^2}$, so that the coordinates of E in base 3 are (.02, 0).

These points are elements of the Cantor Set which is the set of points formed from the closed interval [0, 1] by removing first the middle third of the interval, then the middle third of each remaining interval, and so on indefinitely.

4-10 *Assume that there are* n *stacks of tokens with* n *tokens in each stack. One and only one stack consists entirely of counterfeit tokens, each token weighing* 0.9 *ounce. If each true token weighs* 1.0 *ounce, explain how to identify the counterfeit stack in one weighing, using a scale that gives a reading. You may remove tokens from any stack.*

If all were true tokens, the total weight would be n^2 ounces. Since the counterfeits are lighter, there is an overall deficiency of $\frac{1}{10} n$ ounces. However, thinking in terms of the overall deficiency is not helpful since the situation is unchanged whether the counterfeit stack is the first, the second, the third, . . . , or the n-th.

We must find a way to vary the deficiency in a controllable way!

Label the stacks 1, 2, 3, . . . , n. From the first stack take one token, from the second stack, two tokens, . . . , from the i-th stack, i tokens, $1 \leq i \leq n$. Weigh the collection so obtained.

If all the tokens were true, the weight would be $\frac{1}{2} n(n + 1)$ ounces. (See Appendix VII.) The weight actually obtained will be less than this amount by, say, $\frac{k}{10}$ ounces. This is the key to the solution since a deficiency of $\frac{k}{10}$ ounces implies k counterfeit tokens. Since k tokens came from the k-th stack, it is the counterfeit stack.

For example, if the number of stacks is 10, we weigh $1 + 2 + \cdots + 10 = \frac{1}{2}(10)(11) = 55$ tokens. If they were all true, the weight would be 55 ounces. Let us say that the deficiency is $\frac{1}{2}$ ounce. There are, therefore, 5 counterfeit tokens since $k \cdot \frac{1}{10} = \frac{1}{2}$ implies that $k = 5$. The counterfeit stack is the one numbered 5.

Challenge 3 *Solve the generalized problem of* n *stacks with* n *tokens each, if each true token weighs* t *ounces and each counterfeit weighs* s *ounces. Then apply the result to Problem 4-10 and its challenges.*

For a deficiency of r ounces, $t > s$, the counterfeit stack is $\frac{r}{t - s}$. For an excess of r ounces, $t < s$, the counterfeit stack is $\frac{r}{s - t}$. A single answer for both cases is $\frac{r}{|t - s|}$. How do you interpret a non-integral value of $\frac{r}{|t - s|}$?

5   Equations, Inequations, and Pitfalls

5-1 *Find the solution set of the equation $\dfrac{2x}{x-2} = \dfrac{4}{x-2}$.*

A formal procedure yields $2x = 4$, $x = 2$. But $x = 2$ is an unusable result since it leads to division by zero.

EXPLANATION: To obtain a root of an equation the steps in the solution must be *reversible*. Put another way, each succeeding equation in the solution must be equivalent to the preceding one; that is, the manipulations must produce logically equivalent sentences since logically equivalent sentences define the same set.

From $x = 2$ it is permissible to go to $2x = 4$, but to go from $2x = 4$ to $\dfrac{2x}{x-2} = \dfrac{4}{x-2}$ is not permissible since it involves division by zero. Therefore, the solution set is the null set.

5-2 *Find the pairs of numbers* x, y *such that* $\dfrac{x-3}{2y-7} = x - 3$.

For $2y - 7 \neq 0$, $x - 3 = (x - 3)(2y - 7)$ $\therefore (x - 3) \times (2y - 7 - 1) = 0$. For any real value of y, except $3\frac{1}{2}$, $x = 3$, and for any real value of x, $y = 4$.

5-3 *Find all the real values of* x *such that* $|\sqrt{x} - \sqrt{2}| < 1$.

Since $-1 < \sqrt{x} - \sqrt{2} < 1$, $\sqrt{2} - 1 < \sqrt{x} < \sqrt{2} + 1$, we have, upon squaring the inequalities, $3 - 2\sqrt{2} < x < 3 + 2\sqrt{2}$.

Therefore, x can have any value between 0.172 (approx.) and 5.828 (approx.).

Challenge   *Let the set of all values of* x *satisfying the inequalities* $|x - 8| < 6$ *and* $|x - 3| > 5$ *be written as* a < x < b. *Find* b − a.

Since $|x - 8| < 6$, $x - 8 < 6$, and $x - 8 > -6 \therefore x < 14$, and $x > 2$.

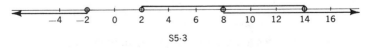

S5-3

Similarly, from $|x - 3| > 5$ we find $x < -2$, or $x > 8$. The overlap of these intervals is $8 < x < 14$, $\therefore b - a = 6$.

In Fig. S5-3, the values of x satisfying the inequalities $|x - 8| < 6$ are shown by the solid line; those satisfying the inequalities $|x - 3| > 5$, by the two arrows. The overlap consists of x-values greater than 8 and less than 14.

5-4 *Find all values of* x *satisfying the equation* $2x = |x| + 1$.

Since the right side of the equation is positive for all values of x, the left side must also be positive, so that $x > 0$. $\therefore 2x = x + 1$, $x = 1$.

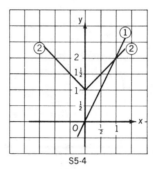

S5-4

It is instructive to look at a geometric interpretation of this equation, as shown in Fig. S5-4. The graph of $y = 2x$ is (1), a line through 0. The graph of $y = |x| + 1$ is (2), the V-shaped broken line. The graphs intersect at point P where $x = 1$.

5-5 *Find values of* a *and* b *so that* $ax + 2 < 3x + b$ *for all* x < 0.

METHOD I: Rearranging, we have $x(a - 3) < b - 2$. Comparing this inequality with $mx < 0$, which holds for all $x < 0$ and all $m > 0$, we conclude that $a - 3 > 0$ and $b - 2 = 0$, so that $a > 3$ and $b = 2$.

METHOD II: Let $x' = -x$ when $x < 0$, $x' > 0$ $\therefore ax' > 3x'$ for $a > 3$. $\therefore ax' + 2 > 3x' + 2$, $a > 3$. Comparing this inequality with the given inequality, we see that $b = 2$, $a > 3$.

5-6 *Find all positive integers that leave a remainder of* 1 *when divided by* 5, *and leave a remainder of* 2 *when divided by* 7.

$N = 5b + 1 = 7a + 2$ so that $b = \dfrac{7a + 1}{5}$ and $a = \dfrac{5b - 1}{7}$.

For b to be an integer a must leave a remainder of 2 when divided by 5; that is, $a = 5k + 2$. Similar reasoning leads to the result

$b = 7L + 3$.   Therefore, $N = 7(5k + 2) + 2 = 35k + 16$, where $k = 0, 1, 2, \ldots$, or $N = 5(7L + 3) + 1 = 35L + 16$, where $L = 0, 1, 2, \ldots$.

Challenge 2   *Solve the problem with the first remainder* $1 \leq r_1 \leq 4$, *and the second remainder* $1 \leq r_2 \leq 6$.

$N = 5b + r_1 = 7a + r_2$ so that $b = \dfrac{7a + r_2 - r_1}{5}$ and $a = \dfrac{5b + r_1 - r_2}{7}$. To obtain integral b, $7a + r_2 - r_1$ must be divisible by 5, so that $a = 5k + 2r_2 - 2r_1$. Similar reasoning leads to the result $b = 7L + 3r_2 - 3r_1$. Therefore,

$N = 7(5k + 2r_2 - 2r_1) + r_2 = 35k + 15r_2 - 14r_1$, or $N = 5(7L + 3r_2 - 3r_1) + r_1 = 35L + 15r_2 - 14r_1$. Verify the solution to the original problem by using this result.

5-7 *On a fence are sparrows and pigeons. When five sparrows leave, there remain two pigeons for every sparrow. Then twenty-five pigeons leave, and there are now three sparrows for every pigeon. Find the original number of sparrows.*

$\dfrac{p}{s - 5} = 2$; $\dfrac{s - 5}{p - 25} = 3$. Solve simultaneously to get $s = 20$ (sparrows) and $p = 30$ (pigeons).

Challenge 1   *Replace "five" by* a *and "twenty-five" by* b, *and find* s *and* p *(the number of sparrows and the number of pigeons, respectively).*

ANSWER: $p = \dfrac{6b}{5}$, $s = a + \dfrac{3b}{5}$. Note that, for this problem to be meaningful, b must be a multiple of 5.

Challenge 2   *Solve the problem generally using* r_1 *and* r_2, *respectively, for the two ratios, and* a *and* b *as in Challenge 1.*

$\dfrac{p}{s - a} = r_1$; $\dfrac{s - a}{p - b} = r_2$

$\therefore \dfrac{p}{p - b} = r_1 r_2$, and $p = \dfrac{r_1 r_2}{r_1 r_2 - 1}$ (b).

$s - a = \dfrac{p}{r_1} = \dfrac{r_2}{r_1 r_2 - 1}$ (b), and $s = a + \dfrac{r_2}{r_1 r_2 - 1}$ (b).

Check the first answers by these formulas.

5-8 *A swimmer at* A, *on one side of a straight-banked canal* 250 *feet wide, swims to a point* B *on the other bank, directly opposite to* A. *His steady rate of swimming is* 3 *ft./sec., and the canal flow is a steady* 2 *ft./sec. Find the shortest time to swim from* A *to* B.

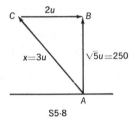

S5-8

Since the swimmer must counteract the effects of the current, he plans his route in terms of vectors, as shown in Fig. S5-8. Since the vector $\overrightarrow{AC}$ plus the vector $\overrightarrow{CB}$ equals the vector $\overrightarrow{AB}$, the swimmer sets his course in the direction of C. But $(AB)^2 = (AC)^2 - (CB)^2 = (3u)^2 - (2u)^2 = 5u^2$ (Pythagorean Theorem). Hence $AB = u\sqrt{5}$. $\therefore \dfrac{x}{250} = \dfrac{3u}{u\sqrt{5}}$, $x = \dfrac{750}{\sqrt{5}}$ (feet).

The shortest time is $\dfrac{750}{\sqrt{5}} \div 3 = \dfrac{250}{\sqrt{5}}$ seconds.

5-9 *Miss Jones buys* x *flowers for* y *dollars, where* x *and* y *are integers. As she is about to leave the clerk says, "If you buy* 18 *flowers more, I can let you have them all for six dollars. In this way you save* 60 *cents per dozen." Find a set of values for* x *and* y *satisfying these conditions.*

$$\frac{100y}{x} - \frac{600}{x + 18} = \frac{60}{12} ; \frac{y}{x} - \frac{6}{x + 18} = \frac{1}{20}$$

$$\frac{1}{20} = \frac{1}{4} - \frac{1}{5} = \frac{3}{12} - \frac{6}{12 + 18} \quad \therefore x = 12,\ y = 3$$

5-10 *Find the set of real values of* x *satisfying the equation*

$$\frac{x + 5}{x + 4} - \frac{x + 6}{x + 5} = \frac{x + 7}{x + 6} - \frac{x + 8}{x + 7}.$$

We note that the equation is in the form

$$\frac{x + a + 1}{x + a} - \frac{x + a + 2}{x + a + 1} = \frac{x + a + 3}{x + a + 2} - \frac{x + a + 4}{x + a + 3}.$$

We can, therefore, solve the problem for an arbitrary value of a with no more difficulty than for the particular value 4.

Combining the two fractions on the left, we obtain $\dfrac{1}{(x + a)(x + a + 1)}$; combining the two on the right, we obtain $\dfrac{1}{(x + a + 2)(x + a + 3)}$. Therefore, $(x + a)(x + a + 1) = (x + a + 2)(x + a + 3)$, and so $x = -\dfrac{2a + 3}{2}$.

For the particular value $a = 4$, $x = -\dfrac{11}{2}$.

COMMENT: In unsimplified form

$$x = -\frac{(a + 2)(a + 3) - a(a + 1)}{[(a + 2) + a + 3] - [a + (a + 1)]}.$$

5-11 *The contents of a purse are not revealed to us, but we are told that there are exactly 6 pennies and at least one nickel and one dime. We are further told that if the number of dimes were changed to the number of nickels, the number of nickels were changed to the number of pennies, and the number of pennies were changed to the number of dimes, the sum would remain unchanged. Find the least possible and the largest possible number of coins the purse contains.*

An obvious solution is 6 dimes, 6 nickels, and 6 pennies. But are 18 coins the least possible? Or the largest possible?

If d and n, respectively, represent the number of dimes and the number of nickels, the condition of constant sums is translated into $10d + 5n + 6 = 5d + n + 60$, with $n \geq 1$, $d \geq 1$. More

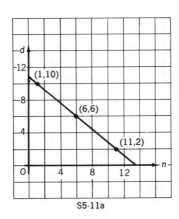

S5-11a

simply, we write $5d = 54 - 4n$ and $d = 10 - \dfrac{4n - 4}{5}$. For d to be a positive integer, $\dfrac{4n - 4}{5}$ must be an integer less than 10. Since $n < 13$, the only acceptable values for n are 11, 6, and 1. The corresponding d-values are 2, 6, and 10. (See Fig. S5-11a.)

The three possibilities are, therefore, 2 dimes, 11 nickels, 6 pennies; 6 dimes, 6 nickels, 6 pennies; 10 dimes, 1 nickel, 6 pennies. The first combination has 19 coins, the largest possible number, and the third combination has 17 coins, the least possible number.

In this instance, the largest combination of coins yields the smallest amount, while the smallest combinations of coins yields the largest amount. We will return to this reversal shortly.

Challenge 1   *How does the situation change if the number of nickels is 6, and the number of dimes and the number of pennies are unspecified, except that there must be at least one of each?*

Here, again, an obvious solution is 6 dimes, 6 nickels, and 6 pennies. But, again, we ask whether 18 coins is the maximum, the minimum, or neither.

$$10d + 30 + c = 5d + 6 + 10c,$$
$$\therefore 5d = 9c - 24, \quad d = \frac{9c - 24}{5}$$

For $\dfrac{9c - 24}{5}$ to be a positive integer, c must have the

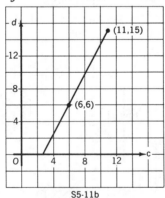

S5-11b

form $5k + 1$, where $k = 1, 2, \ldots$. Acceptable values of c are 6, 11, 16, ..., with 6, 15, 24, ... as the corresponding d-values, as shown in Fig. S5-11b. There are, therefore, an endless number of coin combinations, the first three of which are 6 dimes, 6 nickels, 6 pennies; 15 dimes, 6 nickels, 11 pennies; 24 dimes, 6 nickels, 16 pennies. Here there is no largest combination of coins, and the smallest is the 6-6-6 combination.

Challenge 2   *What solution is obtained if the number of dimes is 6, but the nickels and pennies are unspecified?*

The number of possibilities is endless. The first three are 6 dimes, 6 nickels, 6 cents (the smallest combination); 6 dimes, 15 nickels, 10 cents; and 6 dimes, 24 nickels, 14 cents. There is no largest combination. (See Fig. S5-11c.)

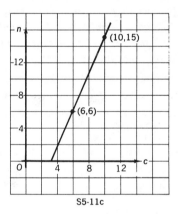

S5-11c

Challenge 3   *Explain why, in the original problem, the least number of coins yields the greatest value, whereas in Challenges 1 and 2 the least number of coins yields the smallest value.*

Consider the original simplified equation $d = 10 - \dfrac{4n - 4}{5}$. By adding $n + 6$ to both sides we have $d + n + 6 = \dfrac{84 + n}{5}$. The left side of this equation represents the total number of coins, N; the relation shows that as n increases on the right, N, too, increases.

By multiplying the equation $d = 10 - \dfrac{4n - 4}{5}$ by 10, and adding to both sides $5n + 6$, we obtain $10d + 5n + 6 = 114 - 3n$. The left side of this equation represents the total value of the coins, V; the relation shows that as n increases on the right, V decreases.

On the other hand, in Challenge 1 we have $d = \frac{9c - 24}{5}$. Therefore, $N = d + 6 + c = \frac{9c - 24}{5} + 6 + c$. As c increases, N increases also. $V = 10d + 30 + c = 19c - 18$, and, as c increases, V increases.

Challenge 4 *Investigate the problem if there are exactly 6 pennies, and at least one nickel, one dime, and one quarter.*

There are many possibilities, one of which is 8 quarters, 4 dimes, 1 nickel, and 6 pennies.

NOTE: The method used in Solution 3-16 may be used in Solution 5-11.

5-12 *A shopper budgets twenty cents for twenty hardware items. Item A is priced at 4 cents each, item B, at 4 for 1 cent, and item C, at 2 for 1 cent. Find all the possible combinations of 20 items made up of items A, B, and C that are purchasable.*

Representing the quantities of items A, B, and C, respectively, by x, y, and z, we translate the given statements into $x + y + z = 20$ and $4x + \frac{1}{4}y + \frac{1}{2}z = 20$. These equations imply that $y = 14x - 40$ and $z = 60 - 15x$.

To avoid fractions of a cent, z must be even and y must be a multiple of 4. But, since $z \le 20$ and $y \ge 0$, x must be 4. However, this value of x makes $z = 0$, and the system of equations is inconsistent.

To obtain a mathematical solution we must allow z to be odd. It follows, then, that $x = 3$, $y = 2$, and $z = 15$. These values satisfy the quantity equation and also the cost equation. From a commercial point of view, however, this may be an unsatisfactory answer.

5-13 *Partition 75 into four positive integers a, b, c, d such that the results are the same when 4 is added to a, subtracted from b, multiplied by c, and divided into d. To partition a positive integer is to represent the integer as a sum of positive integers.*

The given information implies six equalities which have the compact form $a + 4 = b - 4 = 4c = \frac{d}{4}$. Therefore, $b = a + 8$, $c = \frac{a}{4} + 1$, $d = 4a + 16$. Also, since $a + b + c + d = 75$, $a + a + 8 + \frac{a}{4} + 1 + 4a + 16 = 75$. Therefore, $a = 8$, $b = 16$, $c = 3$, $d = 48$.

Challenge 2 *Partition* 100 *into five parts* a, b, c, d, e *so that the results are the same when* 2 *is added to* a, 2 *is subtracted from* b, 2 *is multiplied by* c, 2 *is divided by* d, *and the positive square root is taken of* e.

$$a + 2 = b - 2 = 2c = \frac{d}{2} = \sqrt{e}$$

$$\therefore a + a + 4 + \frac{a}{2} + 1 + 2a + 4 + a^2 + 4a + 4 = 100$$

$$\therefore 2a^2 + 17a - 174 = (2a + 29)(a - 6) = 0$$

$$\therefore a = 6, b = 10, c = 4, d = 16, e = 64$$

Explain the rejection of the result $a = -\dfrac{29}{2}$.

5-14 *Two trains, each traveling uniformly at* 50 *m.p.h., start toward each other, at the same time, from stations* A *and* B, 10 *miles apart. Simultaneously, a bee starts from station* A, *flying parallel to the track at the uniform speed of* 70 *m.p.h., toward the train from station* B. *Upon reaching the train, it comes to rest, and allows itself to be transported back to the point where the trains pass each other. Find the total distance traveled by the bee.*

We designate by X the point where the bee alights upon the train from B. (See Fig. S5-14.) If h designates the fractional part of one

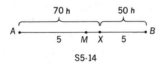

S5-14

hour during which the bee flies, $70h + 50h = 10$, $h = \dfrac{1}{12}$ and $AX = \dfrac{70}{12}$ (miles). The distance it is transported back is $XM = 5 - 50\left(\dfrac{1}{12}\right) = \dfrac{10}{12}$ (miles). The total distance is, therefore, $\dfrac{70}{12} + \dfrac{10}{12} = 6\dfrac{2}{3}$ (miles).

NOTE: M is the midpoint of $\overline{AB}$.

5-15 *One hour out of the station, the locomotive of a freight train develops trouble that slows its speed to* $\dfrac{3}{5}$ *of its average speed up to the time of the failure. Continuing at this reduced speed it reaches its destination two hours late. Had the trouble occurred*

50 miles beyond, the delay would have been reduced by 40 minutes. Find the distance from the station to the destination.

Representing this distance by x (miles) and the normal speed by r (miles per hour), we have, since $T = \dfrac{D}{R}$, normal time $= \dfrac{x}{r}$ and the actual time $= \dfrac{x}{r} + 2$.

Therefore, $1 + \dfrac{x - r}{\frac{3r}{5}} = \dfrac{x}{r} + 2$, and $1 + \dfrac{50}{r} + \dfrac{x - r - 50}{\frac{3r}{5}} = \dfrac{x}{r} + \dfrac{4}{3}$. This pair of equations implies that $x = 200$ (miles). Verify.

5-16 *Two trains, one 350 feet long, the other 450 feet long, on parallel tracks, can pass each other completely in 8 seconds when moving in opposite directions. When moving in the same direction, the faster train completely passes the slower one in 16 seconds. Find the speed of the slower train.*

With the conditions given, is it possible for the trains to have the same speed? Obviously not. So the essential question is, how fast, relatively, are the ends of the trains being separated from each other?

Letting f (feet per second) represent the speed of the faster train, and s (feet per second), the speed of the slower train, the relative speed, when the trains are going in opposite directions, is $f + s$, and the relative speed, when they are going in the same direction, is $f - s$. In either case, the distance traveled is $350 + 450 = 800$ (feet).

Since (relative) rate $\times$ time = distance, we have $(f + s)8 = 800$ and $(f - s)16 = 800$. This pair of equations is easily solved, yielding the values $f = 75$ and $s = 25$ (feet per second).

5-17 *The equation $5(x - 2) = \dfrac{27}{7}(x + 2)$ is written throughout in base 9. Solve for x, expressing its values in base 10.*

METHOD I: We work throughout in base 9. Multiplying both sides of the equation by 7, we have $38(x - 2) = 27(x + 2)$, $38x - 77 = 27x + 55$, $11x = 143$, $x = 13$ (base 9) $\therefore x = 12$ (base 10).

METHOD II: We translate into base 10. Multiplying both sides of the equation by 7, we have $35(x - 2) = 25(x + 2)$, $35x - 70 = 25x + 50$, $10x = 120$, $x = 12$ (base 10).

5-18 *Find the two prime factors of 25,199 if one factor is about twice the other.*

Let the factors be a and b with $a \approx 2b$. Then $ab \approx 2b^2 \approx 25,200$, $b^2 \approx 12,600$, $b \approx 110$. Since b is prime, $b \neq 110$ or 111 or 112. Try $b = 113$. Since $25,199 \div 113 = 223$, the factors are 113 and 223.

Challenge *Find the three prime factors of 27,931 if the three factors are approximately in the ratio 1:2:3.*

Let the factors be a, b, c with $b \approx 2a$ and $c \approx 3a$. Then $6a^3 \approx 27,931$, $a^3 \approx 4655$. Since $16^3 = 4096$ and $17^3 = 4913$, we may take $a = 17$. Since $27,931 \div 17 = 1643$, then $bc \approx \frac{3}{2} b^2 = 1643$, and $b^2 \approx 1095$. Therefore, $b \approx 33$, and since $b \approx 2a$, the choice for b is narrowed to 31 or 37. Of these two, 31 is an exact divisor of 1643, with a quotient 53, while 37 is not. The factors are, therefore, 17, 31, and 53.

5-19 *When asked the time of the day, a problem-posing professor answered, "If you add one-eighth of the time from noon until now to one-quarter the time from now until noon tomorrow, you get the time exactly." What time was it?*

Let h (hours) represent the time interval from noon to now. $\therefore \frac{1}{8} h + \frac{1}{4} (24 - h) = h$, $h = 5\frac{1}{3}$. The time was 5:20 P.M. In essence, this problem is equivalent to the following. If to $\frac{1}{8}$ of a given quantity, you add 6 and then subtract $\frac{1}{4}$ of the quantity, the result equals the original quantity, x. In symbols, $\frac{x}{8} + \frac{1}{4} (24 - x) = x$.

Challenge 1 *On another occasion the professor said, "If from the present time you subtract one-sixth of the time from now until noon tomorrow, you get exactly one-third of the time from noon until now." Find the present time.*

$$h - \frac{1}{6} (24 - h) = \frac{1}{3} h.$$

Time: 4:48 P.M.

5-20 *Solve* $\frac{1}{x} + \frac{1}{y} = \frac{1}{z}$ *for integer values of* x, y, *and* z.

Expressing y in terms of x and z, we have $y = z + \frac{z^2}{x - z}$. For y to be integral $\frac{z^2}{x - z}$ must be integral. Let $w = \frac{z^2}{x - z}$ so that $x = z + \frac{z^2}{w}$, and $y = z + w$. Hence, both x and y will be integers satisfying the given equation if $z = kw$, w an integer, $k = \pm 1, \pm 2, \ldots$. (Does this generate all solutions?)

ILLUSTRATION 1: Let $w = 1$, $k = 1$. Then $x = 2$, $y = 2$, $z = 1$, and $\frac{1}{2} + \frac{1}{2} = 1$.

ILLUSTRATION 2: Let $w = 3$, $k = 5$. Then $x = 90$, $y = 18$, $z = 15$, and $\frac{1}{90} + \frac{1}{18} = \frac{1}{15}$.

ILLUSTRATION 3: Let $w = -2$, $k = 1$. Then $x = -4$, $y = -4$, $z = -2$, and $-\frac{1}{4} - \frac{1}{4} = -\frac{1}{2}$.

ILLUSTRATION 4: Let $w = -5$, $k = -2$. Then $x = -10$, $y = 5$, $z = 10$, and $-\frac{1}{10} + \frac{1}{5} = \frac{1}{10}$.

ILLUSTRATION 5: Investigate the case of $w = -5$, $k = -1$.

5-21 *Prove that, for the same set of integral values of* x *and* y, *both* 3x + y *and* 5x + 6y *are divisible by* 13.

Let $k = 3x + y$, k an integer, so that $x = \frac{k - y}{3}$. Since x is prescribed an integer, $\frac{k - y}{3}$ must be an integer. Let $u = \frac{k - y}{3}$, so that $x = u$ and $y = k - 3u$.

By substitution we have $5x + 6y = 5u + 6k - 18u = 6k - 13u$. Since we are concerned with divisibility by 13, let $k = 13m$. Therefore, $3x + y = 3u + k - 3u = 13m$, and $5x + 6y = 13(6m - u)$, both divisible by 13 for the values $x = u$ and $y = 13m - 3u$ (m is an integer).

Start with the second expression $5x + 6y$ and proceed in an analogous manner to obtain $5x + 6y = 13n$ and $3x + y = 13(-2n + v)$, both divisible by 13 for $x = -13n + 6v$ and $y = 13n - 5v$ (v is an integer)

ILLUSTRATION: Let $u = 3$, $m = 2$. Then $x = 3$, $y = 26 - 9 = 17$. Hence, $3x + y = 9 + 17 = 2 \cdot 13$, and $5x + 6y = 15 + 102 = 9 \cdot 13$.

6 Correspondence: Functionally Speaking

6-1 *Define the symbol* f(a) *to mean the value of a function* f *of a variable* n *when* n = a. *If* f(1) = 1 *and* f(n) = n + f(n − 1) *for all natural numbers* n ≥ 2, *find the value of* f(6).

Rewrite the formula as $f(n) - f(n - 1) = n$. By "telescopic" addition the left side becomes $f(6) - f(1)$. The right side is $2 + 3 + 4 + 5 + 6 = 20$. Therefore, $f(6) = f(1) + 20 = 21$.

$$f(6) - f(5) = 6$$
$$f(5) - f(4) = 5$$
$$f(4) - f(3) = 4$$
$$f(3) - f(2) = 3$$
$$f(2) - f(1) = 2$$

Challenge 2 *If* f(1) = 1 *and* f(n) = n + f(n − 1) *for all natural numbers* n ≥ 2, *show that* f(n) = $\frac{1}{2}$ n (n + 1).

Rewrite the equation $f(n) = n + f(n - 1)$ as $f(n) - f(n - 1) = n$.

$$f(n) - f(n - 1) = n$$
$$f(n - 1) - f(n - 2) = n - 1$$
$$f(n - 2) - f(n - 3) = n - 2$$
$$\vdots$$
$$f(3) - f(2) = 3$$
$$f(2) - f(1) = 2$$

By "telescopic" addition of these $n - 1$ equations, the left side becomes $f(n) - f(1)$. The right side is $2 + 3 + \cdots + n$. Therefore,

$$f(n) = f(1) + 2 + 3 + \cdots + n$$
$$= 1 + 2 + 3 + \cdots + n$$
$$= \frac{1}{2} n (n + 1) \qquad \text{(See Appendix VII.)}$$

When a sequence is defined by giving the initial term (or initial terms) and a formula for finding the successor of any term, we say that the sequence is defined recursively.

6-2 *Each of the following (partial) tables has a function rule associating a value of* n *with its corresponding value* f(n). *If* f(n) = An + B, *determine for each case the numerical values of* A *and* B.

(a) n	f(n)	(b) n	f(n)	(c) n	f(n)	(d) n	f(n)	(e) n	f(n)
$\frac{1}{2}$	$\frac{1}{2}$	$\frac{1}{2}$	$\frac{3}{2}$	$\frac{1}{2}$	0	$2\frac{3}{4}$	$3\frac{1}{2}$	$2\frac{3}{4}$	2
$\frac{1}{3}$	$\frac{2}{3}$	$\frac{1}{3}$	$\frac{4}{3}$	$\frac{1}{3}$	$\frac{1}{3}$	$2\frac{1}{4}$	3	$2\frac{1}{4}$	$1\frac{1}{2}$
$\frac{3}{5}$	$\frac{2}{5}$	$\frac{2}{5}$	$\frac{7}{5}$	$\frac{1}{4}$	$\frac{1}{2}$	1	$1\frac{3}{4}$	1	$\frac{1}{4}$

Some of this can be done by inspection. For example, in **(a)** we notice that $\frac{1}{2} + \frac{1}{2} = 1$, $\frac{1}{3} + \frac{2}{3} = 1$, $\frac{3}{5} + \frac{2}{5} = 1$ ∴ $n + f(n) = 1$ so that $f(n) = 1 - n$, $A = -1$, $B = 1$.

In **(b)** we notice $\frac{3}{2} - \frac{1}{2} = 1$, $\frac{4}{3} - \frac{1}{3} = 1$, $\frac{7}{5} - \frac{2}{5} = 1$ ∴ $f(n) - n = 1$ so that $f(n) = n + 1$, $A = 1$, $B = 1$.

For more difficult cases, a formal procedure may be required. We illustrate for **(c)**. Using values in the table, we determine that

$$0 = \frac{1}{2}A + B, \tag{I}$$

$$\frac{1}{3} = \frac{1}{3}A + B. \tag{II}$$

Subtract equation (II) from equation (I).

$$-\frac{1}{3} = \frac{1}{6}A, \text{ therefore, } A = -2.$$

$$0 = -1 + B; \text{ so } B = 1 \text{ and } f(n) = 1 - 2n.$$

Note that, in this formal procedure, only two sets of values need be used. The third is useful for checking the accuracy of the work. For example, if it appears that $f(n) = 1 - 2n$, Then

$$\frac{1}{2} = 1 - 2\left(\frac{1}{4}\right), \frac{1}{2} = \frac{1}{2} \text{ confirms the validity of the formula.}$$

Do tables **(d)** and **(e)**, either by inspection or by a formal procedure. The answer to **(d)** is $f(n) = n + \frac{3}{4}$, and to **(e)** $f(n) = n - \frac{3}{4}$.

6-3 *In a given right triangle, the perimeter is 30 and the sum of the squares of the sides is 338. Find the lengths of the three sides.*

Letting the lengths of the hypotenuse and the legs be designated by c and a and b, respectively, we have (Pythagorean Theorem)

$c^2 = a^2 + b^2$; thus, $c^2 + b^2 + a^2 = 2c^2 = 338$
$\therefore c^2 = 169$, $c = 13$, and $a^2 + b^2 = 169$.

Using the remaining given information we have

$a + b + c = 30$, $a + b + 13 = 30$, $a + b = 17$, $a = 17 - b$
$\therefore (17 - b)^2 + b^2 = 169$, $b^2 - 17b + 60 = 0$,
$(b - 12)(b - 5) = 0$.

So b is either 12 or 5, and a, accordingly, is either 5 or 12, The sides are, therefore, 5, 12, and 13.

Challenge   *Redo the problem using an area of 30 in place of the perimeter of 30.*

Repeating the first part of the solution above, we have $c = 13$ and $a^2 + b^2 = 169$. Since $\frac{1}{2}ab = 30$, $2ab = 120$. By addition $a^2 + 2ab + b^2 = 289$, and by subtraction $a^2 - 2ab + b^2 = 49$.

So $a + b = 17$, $a - b = 7$ $\therefore a = 12$, $b = 5$.

We could with equal right say, $b + a = 17$, $b - a = 7$ $\therefore b = 12$, $a = 5$. It is immaterial which leg is identified by a.

Why do we reject the values -17 and -7 when taking the square roots of 289 and 49, respectively?

6-4 *A rectangular board is to be constructed to the following specifications:*
(a) the perimeter is equal to or greater than 12 inches, but less than 20 inches
(b) the ratio of adjacent sides is greater than 1 but less than 2.
Find all sets of integral dimensions satisfying these specifications.

METHOD I: Represent the larger and smaller dimensions by x and y, respectively. In Fig. S6-4 are pictured the four inequalities:
(1) $2x + 2y < 20$ or $x + y < 10$ (2) $2x + 2y \geq 12$ or $x + y \geq 6$
(3) $\frac{x}{y} > 1$ (4) $\frac{x}{y} < 2$.

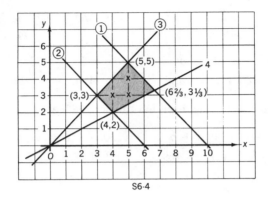

S6-4

The common part of the four regions represented by these inequalities is the shaded quadrilateral.

The required sets are the coordinates of the interior lattice points, namely, (4, 3), (5, 3), and (5, 4). Check these three sets of dimensions against the given information.

METHOD II:

$$\left.\begin{array}{l}\text{Since } x + y < 10, \, y < 10 - x \\ \text{Since } \dfrac{x}{y} > 1, \, y < x\end{array}\right\} \therefore 2y < 10 \text{ and } y < 5.$$

$$\left.\begin{array}{l}\text{Since } x + y \geq 6, \, y \geq 6 - x \\ \text{Since } \dfrac{x}{y} < 2, \, 2y > x\end{array}\right\} \therefore 3y > 6 \text{ and } y > 2.$$

Consequently, the possibilities for y are 3 and 4. Since $y < x$ but $2y > x$, the possible values for x are 4 and 5. The acceptable combinations are, therefore, (4, 3), (5, 3), and (5, 4).

6-5 *Find the range of values of* $F = \dfrac{x^2}{1 + x^4}$ *for real values of* x.

F is never negative. (Why?)

The numerator is least when $x = 0$, and when $x = 0$, the value of the denominator is 1. Therefore, $F(\text{minimum}) = 0$.

One way to find $F(\text{maximum})$ is as follows. By dividing numerator and denominator by x^2, $F = \dfrac{x^2}{1 + x^4}$ becomes $F = \dfrac{1}{x^2 + \dfrac{1}{x^2}}$. F will be largest when the denominator $x^2 + \dfrac{1}{x^2}$ is least.

Since $x^2 \cdot \dfrac{1}{x^2} = 1$, the sum $x^2 + \dfrac{1}{x^2}$ is least when $x^2 = \dfrac{1}{x^2}$; that is, when $x = +1$ or -1. (See Appendix III.) For these values of x, $x^2 + \dfrac{1}{x^2} = 2$. Therefore, F(maximum) $= \dfrac{1}{2}$.

The range of values of F is the interval $\left[0, \dfrac{1}{2}\right]$.

Challenge   *Find the largest and the least values of* $f = \dfrac{2x + 3}{x + 2}$ *for* $x \geq 0$.

Compare your results with the following:

$f = \dfrac{2x + 3}{x + 2} = 2 - \dfrac{1}{x + 2}$. When $x = 0$, $f = 1\frac{1}{2}$, the least value. As x grows larger without bound, the fraction $\dfrac{1}{x + 2}$ grows smaller. Therefore, f grows larger, approaching the limiting value 2. A maximum value for f is not achieved.

The value 2 is designated a supremum, a least upper bound.

The situation is pictured in Fig. S6-5.

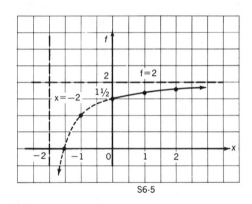

S6-5

6-6 *Determine the largest possible value of the function* $x + 4y$ *subject to the four conditions:* (1) $5x + 6y \leq 30$ (2) $3x + 2y \leq 12$ (3) $x \geq 0$ (4) $y \geq 0$.

Constraints (1), (2), (3), (4), jointly, are represented as the shaded area in which we must find the maximum value of $x + 4y$. See Fig. S6-6. The equation $x + 4y = k$ represents a family of parallel lines, some of which intersect the area in question.

From (1) we have $y \leq 5$ when $x = 0$, and from (2) we have $y \leq 6$ when $x = 0$. Hence, the largest permissible value of y is 5, occurring when $x = 0$.

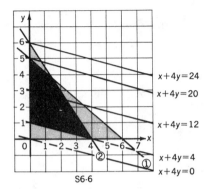

S6-6

It follows that the maximum value of $x + 4y$, subject to the given constraints, is 20.

6-7 *Let us define the distance from the origin* 0 *to point* A *as the length of the* path *along the coordinate lines, as shown in Fig. S6-7, so that the distance from* 0 *to* A *is 3.*

   Starting at 0, *how many points can you reach if the distance, as here defined, is* n, *where* n *is a positive integer?*

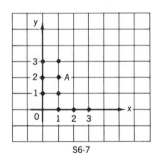

S6-7

When $n = 1$, there are obviously 4 points on the coordinate axes. When $n = 2$, there are 8 points $(2 \cdot 4)$, one in each quadrant and four on the axes.

We can now guess that the answer is $4n$. Let us prove it. When $n = 3$, the 12 points $(3 \cdot 4)$ reached have coordinates $(3, 0)$, $(2, 1)$, $(1, 2)$, $(0, 3)$, and their reflections in the x-axis, the y-axis, and the origin, namely, $(2, -1)$, $(-2, 1)$, $(-2, -1)$, $(1, -2)$, $(-1, 2)$, $(-1, -2)$, and $(-3, 0)$ and $(0, -3)$.

We note that, with respect to integer addition, the number 3 has four partitions: $3 + 0$, $2 + 1$, $1 + 2$, $0 + 3$. These give us the four points (3, 0), (2, 1), (1, 2), (0, 3). Corresponding to (2, 1) are the three reflections (2, −1), (−2, 1), (−2, −1); corresponding to (1, 2) are the three reflections (1, −2), (−1, 2), (−1, −2); corresponding to (3, 0) is the single reflection (−3, 0); and corresponding to (0, 3) is the single reflection (0, −3). The total number of points is $4 + 4 + 2 + 2 = 12$.

Since $4 = 4 + 0 = 3 + 1 = 2 + 2 = 1 + 3 = 0 + 4$, we may guess that the total number of paths is $4 + 4 + 4 + 2 + 2 = 16$. Verify this.

For the general case we have $n = n + 0 = (n − 1) + 1 = (n − 2) + 2 = \cdots = 1 + (n − 1) = 0 + n$, a total of $n + 1$ partitions. The $n − 1$ points, $(n − 1, 1)$, $(n − 2, 2), \ldots$, $(2, n − 2)$, $(1, n − 1)$, together with their reflections (3 each) account for $4n − 4$ paths. Four additional paths are provided by $(n, 0)$ and its reflection, and $(0, n)$ and its reflection. The total is $4(n − 1) + 2 + 2 = 4n$.

You may prefer this alternative view. Since there are $n + 1$ partitions, we are provided with $n + 1$ points of which $n − 1$ points have three reflections each, and two of which have one reflection each. Hence, the total number of paths is $4(n + 1) − 2 − 2 = 4n$, or $4(n − 1) + 2 + 2 = 4n$.

6-8 *Given n straight lines in a plane such that each line is infinite in extent in both directions, no two lines are parallel (fail to meet), and no three lines are concurrent (meet in one point), into how many regions do the n lines separate the plane?*

METHOD I: Let $f(n)$ denote the number of regions into which n such lines separate the plane. By observation, $f(1) = 2$, $f(2) = 4$, $f(3) = 7$.

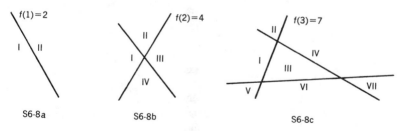

S6-8a   S6-8b   S6-8c

We note that a second line added to Fig. S6-8a splits each of the two regions in two; hence, the second line creates two additional regions. A third line added to Fig. S6-8b splits each of three regions in two and creates three additional regions. Similarly, a fourth line added to Fig. S6-8c splits each of four regions in two, thus adding four regions. Try it!

Let us, therefore, assume that an n-th line added to the diagram with $n - 1$ lines creates n new regions, and proceed to derive, in terms of n, an explicit formula for $f(n)$. By telescopic addition (see Problem 6-1),

$$
\begin{aligned}
f(n) &= f(1) + 2 + 3 + \cdots + n \\
&= 2 + 2 + 3 + \cdots + n \\
&= 1 + (1 + 2 + 3 + \cdots + n) \\
&= 1 + \frac{1}{2} n (n + 1) \qquad \text{(See Appendix VII.)} \\
&= \frac{n^2 + n + 2}{2}
\end{aligned}
$$

Checking, we find that this formula gives us the known values for $n = 1, 2, 3$.

If we now assume that the formula is correct for a natural number $k \geq 3$ (it could just as well be $k \geq 1$), we can show that it is correct for the successor of k, namely, $k + 1$. (See Appendix VII.)

$$
\begin{aligned}
f(k + 1) &= f(k) + k + 1 = \frac{k^2 + k + 2}{2} + k + 1 \\
&= \frac{k^2 + k + 2 + 2k + 2}{2} = \frac{(k + 1)^2 + (k + 1) + 2}{2}
\end{aligned}
$$

Since $k + 1$ is the successor of k (where $k \geq 1$), the formula holds for all natural numbers.

METHOD II: By observation we obtain $f(1) = 2$, $f(2) = 4$, $f(3) = 7, f(4) = 11$.

	$n = 1$	$n = 2$	$n = 3$	$n = 4$
$f(n)$	2	4	7	11
$\Delta f(n)$		2	3	4
$\Delta^2 f(n)$			1	1	(Constant)

The line $\Delta f(n)$ shows the first differences derived from the line $f(n)$, and the line $\Delta^2 f(n)$ shows the first differences derived from the line $\Delta f(n)$, or the second differences derived from the line $f(n)$.

Since the second differences are constant we represent $f(n)$ by a third-degree polynomial $f(n) = An^3 + Bn^2 + Cn + D$, with the numerical values of A, B, C, and D to be determined.

When $n = 1$ we have $\quad 2 = \quad A + \quad B + \quad C + D$,

$\qquad n = 2$ we have $\quad 4 = \quad 8A + \quad 4B + 2C + D$,

$\qquad n = 3$ we have $\quad 7 = 27A + \quad 9B + 3C + D$,

$\qquad n = 4$ we have $11 = 64A + 16B + 4C + D$.

The solution of this system yields $A = 0, B = \dfrac{1}{2}, C = \dfrac{1}{2}, D = 1$.

$$\therefore f(n) = \frac{1}{2}n^2 + \frac{1}{2}n + 1 = \frac{n^2 + n + 2}{2}$$

Challenge 1 *Let there be* $n = r + k$ *lines in the plane (infinite in both directions) such that no three of the* n *lines are concurrent, but* k *lines are parallel (but no others). Find the number of partitions of the plane.*

Let $f(r, k)$ denote the number of separate regions. Since the k-th parallel crosses r lines, it creates $r + 1$ new regions, so that $f(r, k) = f(r, k - 1) + r + 1$. Using telescopic addition on the k equations,

$$f(r, k) - f(r, k - 1) = r + 1$$
$$f(r, k - 1) - f(r, k - 2) = r + 1$$
$$f(r, k - 2) - f(r, k - 3) = r + 1$$
$$\vdots$$
$$f(r, 2) - f(r, 1) = r + 1$$
$$f(r, 1) - f(r, 0) = r + 1.$$

We obtain $f(r, k) - f(r, 0) = k(r + 1)$, so that

$$f(r, k) = f(r, 0) + k(r + 1).$$

But $f(r, 0) = f(r) = \dfrac{r^2 + r + 2}{2}$;

therefore, $f(r, k) = \dfrac{r^2 + r + 2}{2} + k(r + 1)$.

Challenge 2 *Let there be* n *straight lines in the plane (infinite in both directions) such that three (and only three) are concurrent and such that no two are parallél. Find the number of plane separations.*

Let $f(n-3, 3)$ denote the number of regions of separation. Consider a typical region R bounded by three lines, illustrated in Fig. S6-8d. Let l_2 and l_3 meet in point P. Translate or rotate l_1 so that it goes through P; the region R is eliminated.

$$\therefore f(n-3, 3) = \frac{n^2 + n + 2}{2} - 1 = \frac{1}{2} n(n+1)$$

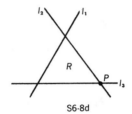

S6-8d

6-9 *Define* $\dfrac{N}{D}$ *as a proper fraction when* $\dfrac{N}{D} < 1$ *with* **N, D** *natural numbers. Let* f(D) *be the number of irreducible proper fractions with denominator* D. *Find* f(D) *for* D $= 51$.

First we note that, at most, there are 50 possible numerators beginning with 1 and ending with 50.

Now we must remove the reducible cases. Since $51 = 3 \cdot 17$, then $3k < 51$, $k \le 16$, so that there are 16 reducible fractions with multiples of 3 in the numerator. Similarly, $17L < 51, L \le 2$, so that there are 2 reducible fractions with multiples of 17 in the numerator.

$$\therefore f(D) = (51 - 1) - (16 + 2) = 32$$

Challenge *Find* f(D) *for* D $= 52$.

Since $52 = 2^2 \cdot 13$, therefore, $13k < 52, k \le 3, 2L < 52$, and $L \le 25$. However, the case $26 = 2 \cdot 13$ is duplicated, $\therefore f(D) = (52 - 1) + (25 + 3 - 1) = 24$.

7   Equations and Inequations: Traveling in Groups

7-1 *Let the lines* $15x + 20y = -2$ *and* $x - y = -2$ *intersect in point* P. *Find all values of* k *which ensure that the line* $2x + 3y = k^2$ *goes through point* P.

The solution set of the pair of equations $15x + 20y = -2$ and $x - y = -2$ is $\left\{ x = -\dfrac{6}{5},\ y = \dfrac{4}{5} \right\}$. Therefore, $2x + 3y = k^2$ implies that $2\left(-\dfrac{6}{5}\right) + 3\left(\dfrac{4}{5}\right) = k^2$ so that $k^2 = 0$ and, hence, $k = 0$.

7-2 *Let* (x, y) *be the coordinates of point* P *in the xy-plane, and let* (X, Y) *be the coordinates of point* Q *(the image of point* P*) in the XY-plane. If* $X = x + y$ *and* $Y = x - y$, *find the simplest equation for the set of points in the XY-plane which is the image of the set of points* $x^2 + y^2 = 1$ *in the xy-plane.*

Since $X = x + y$, $X^2 = x^2 + 2xy + y^2$; and since $Y = x - y$, $Y^2 = x^2 - 2xy + y^2$. Therefore, $X^2 + Y^2 = 2(x^2 + y^2)$. Since $x^2 + y^2 = 1$, $2(x^2 + y^2) = 2$. Hence, $x^2 + y^2 = 2$.

In geometric terms, the image set is a circle with center at $(0, 0)$ and a radius of length $\sqrt{2}$.

7-3 *The numerator and the denominator of a fraction are integers differing by* 16. *Find the fraction if its value is more than* $\dfrac{5}{9}$ *but less than* $\dfrac{4}{7}$.

Let n represent the numerator of the fraction F. Since $F < 1$, the denominator is greater than the numerator, and so $n + 16$ represents the denominator.

Since $\dfrac{5}{9} < \dfrac{n}{n + 16}$, $5n + 80 < 9n$, $80 < 4n$, and so $20 < n$.

Since, also, $\dfrac{n}{n + 16} < \dfrac{4}{7}$, $7n < 4n + 64$, $3n < 64$, and $n < 21\dfrac{1}{3}$.

Since n is an integer and $20 < n < 21\dfrac{1}{3}$, $n = 21$.

It follows that $F = \dfrac{21}{37}$.

7-4 *If* $x + y + 2z = k$, $x + 2y + z = k$, *and* $2x + y + z = k$, $k \neq 0$, *express* $x^2 + y^2 + z^2$ *in terms of* k.

Addition of the three equations yields $4(x + y + z) = 3k$.
$\therefore x + y + z = \dfrac{3k}{4}$;

$$(x + y + z)^2 = x^2 + y^2 + z^2 + 2(xy + yz + zx) = \frac{9k^2}{16}$$

Subtraction applied to the given equations in pairs yields $y - z = 0$, $x - y = 0$, $x - z = 0$. Therefore, $y^2 + z^2 = 2yz$, $x^2 + y^2 = 2xy$, $x^2 + z^2 = 2zx$, so that $2(xy + yz + zx) = 2(x^2 + y^2 + z^2)$. $\therefore 3(x^2 + y^2 + z^2) = \dfrac{9k^2}{16}$, and $x^2 + y^2 + z^2 = \dfrac{3k^2}{16}$.

Challenge 1 Solve the problem if $x + 2y + 3z = k$, $3x + y + 2z = k$, and $2x + 3y + z = k$, $k \neq 0$.

Addition of the three equations yields:

$$6(x + y + z) = 3k, \text{ and } x + y + z = \frac{k}{2}.$$

$$(x + y + z)^2 = x^2 + y^2 + z^2 + 2(xy + yz + xz) = \frac{k^2}{4}.$$

Subtracting the first given equation from the second, we have $2x - y - z = 0$ so that $y + z = 2x$. Subtracting the second given equation from the third, we have $-x + 2y - z = 0$ so that $z + x = 2y$. Subtracting the third given equation from the first, we have $-x - y + 2z = 0$ so that $x + y = 2z$. Hence, $y^2 + 2yz + z^2 = 4x^2$, $z^2 + 2zx + x^2 = 4y^2$, and $x^2 + 2xy + y^2 = 4z^2$. Therefore, $2(xy + yz + zx) = 2(x^2 + y^2 + z^2)$, $\therefore 3(x^2 + y^2 + z^2) = \dfrac{k^2}{4}$ so that $x^2 + y^2 + z^2 = \dfrac{k^2}{12}$.

Challenge 2 Show that both the problem and Challenge 1 can be solved by inspection.

PROBLEM: System satisfied for $x = y = z = \dfrac{k}{4}$
$\therefore x^2 + y^2 + z^2 = \dfrac{3k^2}{16}$.

CHALLENGE: System satisfied for $x = y = z = \dfrac{k}{6}$
$\therefore x^2 + y^2 + z^2 = \dfrac{k^2}{12}$.

7-5 *Why are there no integer solutions of* $x^2 - 5y = 27$?

Rewrite the equation as $x^2 = 5y + 27 = 5(y + 5) + 2 = 5z + 2$, where $z = y + 5$. The integer x can take one of the forms $5m$, $5m + 1$, $5m + 2$, $5m - 1$, $5m - 2$.

CASE I: $x^2 = 5z + 2 \neq 25m^2$ since, upon division by 5, the left side yields the remainder 2 while the right side yields the remainder 0.

CASES II AND IV: $x^2 = 5z + 2 \neq 25m^2 \pm 10m + 1$, since then the respective remainders are 2 and 1.

CASES III AND V: $x^2 = 5z + 2 \neq 25m^2 \pm 20m + 4$, since then the respective remainders are 2 and 4.

7-6 *Civic Town has 500 voters, all of whom vote on two issues in a referendum. The first issue shows 375 in favor, and the second issue shows 275 in favor. If there are exactly 40 votes against both issues, find the number of votes in favor of both issues.*

METHOD I: The number of votes against either issue is $125 + 225 - 40 = 310$. (Explain the subtraction of 40.) Therefore, $500 - 310 = 190$ votes were cast in favor of both issues.

METHOD II: Let x be the number of voters in favor of both issues, let y be the number of voters in favor of the first issue but against the second, and let z be the number of voters in favor of the second issue but against the first. In Fig. S7-6 the information is shown in tabular form.

		275	225
I \ II		For	Against
375	For	x	y
125	Against	z	$w = 40$

$x + y = 375$, $x + z = 275$, $y + 40 = 225$, $z + 40 = 125$

Therefore, $y = 185$, $z = 85$, $x = 190$ (number in favor of both issues).

7-7 *How do you find the true weight of an article on a balance scale in which the two arms (distances from the pans to the point of support) are unequal?*

Let x(pounds) be the weight of the article, d_1(inches) the length of the right arm, and d_2(inches) the length of the left arm.

Place the article in the right pan and let it be balanced by a weight w_1(pounds) in the left pan. Then $xd_1 = w_1d_2$(Principle of Moments). Place the article in the left pan and let it be balanced by a weight w_2(pounds) in the right pan. Then $xd_2 = w_2d_1$.

Therefore, $x^2d_1d_2 = w_1w_2d_1d_2$, $x^2 = w_1w_2$, $x = \sqrt{w_1w_2}$; that is, the true weight of the article is the geometric mean of the two weights used. (See Appendix IV.)

Challenge *Suppose it is known that the arms of a balance scale are unequal; how do you determine the ratio* r *of the arm-lengths?*

Place in the right pan a known weight; for convenience, choose a 1-lb. weight. Find the weight w_1(pounds) needed to produce balance. Therefore, $1d_1 = w_1d_2$. Now, place the 1-lb. weight in the left pan, and let w_2(pounds) produce balance. Therefore, $1d_2 = w_2d_1$.

We have $w_1 = \dfrac{d_1}{d_2}$, $w_2 = \dfrac{d_2}{d_1}$, $w_1 + w_2 = \dfrac{d_1}{d_2} + \dfrac{d_2}{d_1} = r + \dfrac{1}{r}$.

If $w_1 + w_2$ is known, $r \left(= \dfrac{d_1}{d_2} \right)$ is determined.

ILLUSTRATION: If $w_1 + w_2 = \dfrac{61}{30}$ (pounds), then $r + \dfrac{1}{r} = \dfrac{61}{30}$, $r^2 - \dfrac{61r}{30} + 1 = 0$, $\left(r - \dfrac{5}{6} \right)\left(r - \dfrac{6}{5} \right) = 0$, $r = \dfrac{5}{6}$ or $r = \dfrac{6}{5}$.

7-8 *Solve in base 7 the pair of equations* $2x - 4y = 33$ *and* $3x + y = 31$, *where* x, y, *and the coefficients are in base 7.*

METHOD I: (working in base 7)

$2x - 4y = 33$		(I)
$3x + y = 31$		(II)
$15x + 4y = 154$	(4 times II)	(III)
$20x = 220$	(I + III)	(IV)
$x = 11, y = -2$		

METHOD II: (working in base 10)

$$2x - 4y = 24 \tag{I}$$

$$3x + y = 22 \tag{II}$$

$$12x + 4y = 88 \qquad \text{(4 times II)} \tag{III}$$

$$14x = 112 \qquad \text{(I + III)} \tag{IV}$$

$x = 8$ (base 10) $= 11$ (base 7)

$y = -2$ (base 10) $= -2$ (base 7)

7-9 *Given the pair of equations* $2x - 3y = 13$ *and* $3x + 2y = b$, *where* b *is an integer and* $1 \leq b \leq 100$, *let* $n^2 = x + y$, *where* x *and* y *are integer solutions of the given equations associated in proper pairs. Find the positive values of* n *for which these conditions are met.*

Eliminate y from the given equations to obtain $13x = 26 + 3b$, or $x = 2 + \frac{3b}{13}$. Eliminate x from the given equations to obtain $13y = -39 + 2b$, or $y = -3 + \frac{2b}{13}$. Therefore, $x + y = \frac{5b}{13} - 1$. When $b = 13$, $n^2 = x + y = 4$, and so $n = 2$. When $b = 26$, $n^2 = x + y = 9$, and so $n = 3$.

For the remaining multiples of 13 between 1 and 100, we fail to obtain for $x + y$ the square of an integer. There are, consequently, just two values of n, namely, $n = 2$ and $n = 3$.

7-10 *Find the set of integer pairs satisfying the system*

$$3x + 4y = 32$$
$$y > x$$
$$y < \frac{3}{2}x.$$

METHOD I: (Algebraic) Since $y > x, 4y > 4x$. $\qquad$ (I)

Since $3x + 4y = 32$, $\qquad$ (II)

$-3x > 4x - 32.$ $\qquad$ (I) $-$ (II)

Therefore, $7x < 32$, $x < \frac{32}{7}$, and $x \leq 4$.

Since $y < \frac{3}{2}x, 4y < 6x$, and $\qquad$ (III)

$-3x < 6x - 32,$ $\qquad$ (III) $-$ (II)

$9x > 32, x > \frac{32}{9} > 3.$

$3 < x \leq 4$ so that $x = 4$ and, consequently, $y = 5$.

Check $y(=5) > x(=4)$, and $y(=5) < \frac{3}{2}x(=6)$.

Also $3x + 4y = 32$, since $3 \cdot 4 + 4 \cdot 5 = 32$.

METHOD II: (Geometric) The required integral values are on the segment $\overline{P_2P_1}$ of the line $3x + 4y = 32$, as shown in Fig. S7-10. The only set of such integral values are $x = 4$ and $y = 5$.

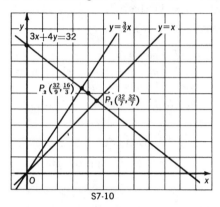

S7-10

7-11 *Compare the solution of system I,*

$$x + y + z = 1.5 \tag{1}$$
$$2x - y + z = 0.8 \tag{2}$$
$$x - 2y - z = -0.2, \tag{3}$$

with that of system II,

$$x + y + z = 1.5 \tag{1}$$
$$2x - y + z = 0.9 \tag{2}$$
$$x - 2y - z = -0.2. \tag{3}$$

A small change in the conditions of a problem can make a big difference in the result!

The systems may be solved by any one of a number of methods. By the Method of Elimination, we obtain, for System I,

$2x - y = 1.3$, by adding equations (1) and (3), and

$3x - 3y = 0.6$, by adding equations (2) and (3).

This new system yields $x = 1.1$, and $y = 0.9$. From either of the original equations we obtain $z = -0.5$.

For System II, a similar procedure yields $x = \frac{32}{30}$, $y = \frac{25}{30}$, $z = -\frac{12}{30}$. Therefore, the change in x is $\frac{33}{30} - \frac{32}{30} = \frac{1}{30}$, the change in y is $\frac{27}{30} - \frac{25}{30} = \frac{2}{30}$, and the change in z is $-\frac{15}{30} - \left(-\frac{12}{30}\right) = -\frac{3}{30}$.

7-12 *The following information was obtained by measurement in a series of experiments:*

$$x + y = 1.9$$
$$2x - y = 1.4$$
$$x - 2y = -0.6$$
$$x - y = 0.3.$$

Find an approximate solution to this system of equations.

From a mathematical viewpoint, the system is inconsistent. However, we cannot reasonably expect "exactness" in measurements, and so it is meaningful to seek an approximate solution to the system.

We consider that each of the constants on the right side of the equations is "in error" by a certain amount, and it is our purpose to obtain a solution so that the errors are minimized. There is more than one way to do this; the method used here is known as the Method of the Mean.

Starting with the assumption that the error in one equation is neither more nor less weighty than the error in any other equation, we group the equations, making as many groups as there are unknowns, in this case, two. Group I consists of the first and second equations, and Group II, of the third and fourth equations. By addition within each group, we obtain:

$$3x = 3.3$$
$$2x - 3y = -0.3.$$

The solution to this system is $x = \frac{11}{10}$, $y = \frac{5}{6}$.

These values satisfy none of the equations exactly, but the sum of the errors is minimized to the value zero, since:

(1) $x + y = \frac{58}{30}$ with an error of $\frac{58}{30} - \frac{57}{30} = +\frac{1}{30}$

(2) $2x - y = \dfrac{41}{30}$ with an error of $\dfrac{41}{30} - \dfrac{42}{30} = -\dfrac{1}{30}$

(3) $x - 2y = -\dfrac{17}{30}$ with an error of $-\dfrac{17}{30} + \dfrac{18}{30} = +\dfrac{1}{30}$

(4) $x - y = \dfrac{8}{30}$ with an error of $\dfrac{8}{30} - \dfrac{9}{30} = -\dfrac{1}{30}$.

7-13 *Find the maximum and the minimum values of the function* $3x - y + 5$, *subject to the restrictions* $y \geq 1$, $x \leq y$, *and* $2x - 3y + 6 \geq 0$.

The conditions stated in the problem confine the values of the given function f: $3x - y + 5$ to the values of x and y determined by the triangle $V_1 V_2 V_3$ obtained by the intersections of $l_1(y = x)$, $l_2(y = 1)$, and $l_3(2x - 3y + 6) = 0$. See Fig. S7-13a.

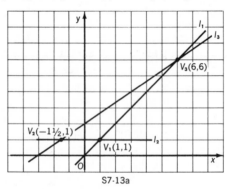

S7-13a

At V_1 (1, 1), the value of the function f is $3 - 1 + 5 = 7$; at V_2 $(-1\frac{1}{2}, 1)$, the value of f is $-4\frac{1}{2} - 1 + 5 = -\frac{1}{2}$; and at V_3 (6, 6), the value of f is $18 - 6 + 5 = 17$. At interior points of the triangle, the value of f is greater than $-\frac{1}{2}$ and less than 17. (See proof below.) Therefore, subject to the conditions given, the maximum value of $3x - y + 5$ is 17, and the minimum value of $3x - y + 5$ is $-\frac{1}{2}$.

The Linear Program Theorem states that the maximum and minimum values of the linear function $f(x, y) = Ax + By + C$ occur at vertices of the polygon S where

$$S = F_1 \cap F_2 \cap \cdots \cap F_n \text{ and } F_n = \{P \mid A_n x + B_n y + C_n \geq 0\}.$$

In Problem 7-13 the polygon is a triangle; in Problem 7-14 the polygon is a hexagon.

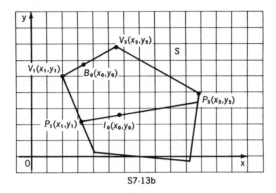

S7-13b

PROOF: Consider an interior point I_0 (x_0, y_0) (Fig. S7-13b). Then $F_1(x_0, y_0) > 0$, $F_2(x_0, y_0) > 0, \ldots, F_n(x_0, y_0) > 0$. Let $\overleftrightarrow{P_1P_2}$ be a line through I_0, intersecting two sides of polygon S in P_1 and P_2 and not parallel to $f(x, y) = 0$. Since $I_0 \in \overleftrightarrow{P_1P_2}$, then $f(x_1, y_1) \neq f(x_2, y_2)$, and either $f(x_1, y_1) < f(x_0, y_0) < f(x_2, y_2)$, or $f(x_2, y_2) < f(x_0, y_0) < f(x_1, y_1)$. Therefore, $f(x_0, y_0)$ is neither a maximum nor a minimum value of $f(x, y)$.

Now consider a point, $B_0(x_0, y_0)$, interior to a side of S with vertices $V_1(x_1, y_1)$ and $V_2(x_2, y_2)$. Then, if $f(x_1, y_1) \neq f(x_2, y_2)$, either $f(x_1, y_1) < f(x_0, y_0) < f(x_2, y_2)$, or $f(x_2, y_2) < f(x_0, y_0) < f(x_1, y_1)$, and if $f(x_1, y_1) = f(x_2, y_2), f(x_1, y_1) = f(x_0, y_0) = f(x_2, y_2)$.

Among the vertices, therefore, there is one such that the value of $f(x, y)$ is at least as great as its value at any other point on the polygon and greater than the value at any interior point. Therefore, the maximum value of $f(x, y)$ occurs at one or more vertices.

A similar argument holds for minimum value.

7-14 *A buyer wishes to order* 100 *articles of three types of merchandise identified as* A, B, *and* C, *each costing* $5, $6, *and* $7, *respectively. From past experience, he knows that the number of each article bought should not be less than* 10 *nor more than* 60, *and that the number of* B *articles should not exceed the number of* A *articles by more than* 30. *If the selling prices for the articles are* $10 *for*

A, \$15 *for* B, *and* \$20 *for* C, *and all the articles are sold, find the number of each article to be bought so that there is a maximum profit.*

The conditions of the problem lead to the following statements, illustrated by Fig. S7-14.

(1) $x + y + z = 100$ so that $z = 100 - x - y$

(2) Cost $= 5x + 6y + 7z = 700 - 2x - y$

(3) $10 \leq x \leq 60$, $10 \leq y \leq 60$, $10 \leq z \leq 60$ so that $10 \leq 100 - x - y \leq 60$ and, therefore, $40 \leq x + y \leq 90$

(4) $y \leq x + 30$

(5) Sales $= 10x + 15y + 20z = 2000 - 10x - 5y$

(6) Gain $=$ Sales $-$ Cost $= 1300 - 8x - 4y$, the function to be maximized

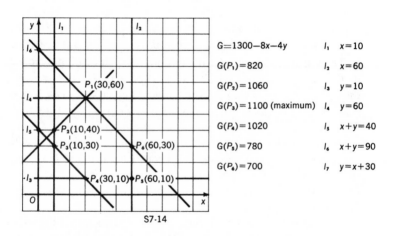

$G = 1300 - 8x - 4y$	$l_1 \quad x = 10$
$G(P_1) = 820$	$l_2 \quad x = 60$
$G(P_2) = 1060$	$l_3 \quad y = 10$
$G(P_3) = 1100$ (maximum)	$l_4 \quad y = 60$
$G(P_4) = 1020$	$l_5 \quad x + y = 40$
$G(P_5) = 780$	$l_6 \quad x + y = 90$
$G(P_6) = 700$	$l_7 \quad y = x + 30$

S7-14

For maximum profit, then, the distribution of the 100 items bought should be 10 item A, 30 item B, and 60 item C. (See the solution to Problem 7-13.)

8   Miscellaneous: Curiosity Cases

8-1 *Find all values of* x *satisfying the equation* $\dfrac{x - \sqrt{x+1}}{x + \sqrt{x+1}} = \dfrac{11}{5}$.

METHOD I: By multiplication obtain $5x - 5\sqrt{x+1} = 11x + 11\sqrt{x+1}$. Therefore, $-3x = 8\sqrt{x+1}$, $9x^2 = 64x + 64$, $9x^2 - 64x - 64 = 0$. Since $9x^2 - 64x - 64 = (9x + 8) \times (x - 8)$, either $9x + 8 = 0$ or $x - 8 = 0$. Therefore, $x = -\dfrac{8}{9}$ ($x = 8$ is rejected).

METHOD II: Since $\dfrac{x - \sqrt{x+1}}{x + \sqrt{x+1}} = \dfrac{11}{5}$, we may write $\dfrac{x - \sqrt{x+1}}{x + \sqrt{x+1}} = \dfrac{-11}{-5}$, and so

$$x - \sqrt{x+1} = -11k, \text{ and} \qquad \text{(I)}$$
$$x + \sqrt{x+1} = -5k, \qquad \text{(II)}$$

where k is a positive constant. By adding (I) and (II) we have $2x = -16k$, and so $x = -8k$. $-8k + \sqrt{1 - 8k} = -5k$; that is, $\sqrt{1 - 8k} = 3k$ so that $1 - 8k = 9k^2$, and $9k^2 + 8k - 1 = 0$. But $9k^2 + 8k - 1 = (9k - 1)(k + 1)$. Therefore, $9k - 1 = 0$ and $k = \dfrac{1}{9}$ ($k = -1$ is rejected). Hence, $x = -\dfrac{8}{9}$.

8-2 *Find all real values of* x *satisfying the equations:*
(a) $x^2|x| = 8$
(b) $x|x^2| = 8$, *where the symbol* |x| *means* +x *when* x $\geq$ 0, *and* −x *when* x < 0.

(a) When $x > 0$, $(x^2|x| = 8) \Rightarrow (x^3 = 8)$ so that $x = 2$. (The symbol $\Rightarrow$ is read "implies".) When $x < 0$, $(x^2|x| = 8) \Rightarrow (-x^3 = 8)$ or $(x^3 = -8)$ so that $x = -2$.
(b) When $x > 0$, $(x|x^2| = 8) \Rightarrow (x^3 = 8)$ so that $x = 2$. Since $8 > 0$, x cannot be negative.
Besides these real values of x, there are imaginary values of x.

8-3 *Let* P(x, y) *be a point on the graph of* y = x + 5. *Connect* P *with* Q(7, 0). *Let a perpendicular from* P *to the* x-*axis intersect it in* R. *Restricting the abscissa of* P *to values between* 0 *and* 7, *both included, find:*
(a) *the maximum area of right triangle* PRQ
(b) *two positions of* P *yielding equal areas.*

(a) Designate the area of $\triangle PRQ$ by K (expressed in proper units). Then $K = \frac{1}{2}(7 - x)(x + 5) = \frac{1}{2}[36 - (x - 1)^2]$. The expression $\frac{1}{2}[36 - (x - 1)^2]$ is a maximum when $x = 1$ for $0 \leq x \leq 7$. $\therefore K(\max) = \frac{1}{2}(6)(6) = 18$. An alternative solution for (a): The area of $\triangle PRQ$ is determined by the measure of the segments $\overline{RP}$ and $\overline{RQ}$. At all times, the length of $\overline{RP}$ is $x + 5$ and the length of $\overline{RQ}$ is $7 - x$. Their sum is the constant 12. It is known (see Appendix III) that when the sum of two positive numbers is fixed, the maximum product of the two numbers occurs when each is one-half the fixed sum. Therefore, the maximum area is obtained when $RP = RQ = 6$, so that the maximum area is $\frac{1}{2}(6)(6) = 18$. The triangle with maximum area is isosceles.

(b) $K_1 = \frac{1}{2}(7 - x_1)(x_1 + 5)$, $K_2 = \frac{1}{2}(7 - x_2)(x_2 + 5)$, and $K_1 = K_2$.

Therefore, $7x_1 + 35 - x_1^2 - 5x_1 = 7x_2 + 35 - x_2^2 - 5x_2$, and so, $2(x_1 - x_2) = x_1^2 - x_2^2 = (x_1 + y_2)(x_1 - y_2)$.

Since we are assuming that $x_1 \neq x_2$, we have $2 = x_1 + x_2$ (dividing by $x_1 - x_2$). One possibility is $x_1 = 0, x_2 = 2$, so that $P_1(0, 5)$ and $P_2(2, 7)$. A second possibility is $P_1\left(\frac{1}{2}, 5\frac{1}{2}\right)$, $P_2\left(\frac{3}{2}, 6\frac{1}{2}\right)$. There are others, of course.

8-4 *Find the smallest value of* x *satisfying the conditions:* $x^3 + 2x^2 = a$, *where* x *is an odd integer, and* a *is the square of an integer.*

$x^3 + 2x^2 = x^2(x + 2) = m^2n^2$ where m, n are integers. $\therefore x + 2 = m^2$ or n^2. The smallest odd value of x, such that $x + 2$ is the square of an integer, is $x = 7$, since $x = 1$ or 3 or 5 are unacceptable.

If we remove the restriction, "smallest value of x," there are, of course, an endless number of x-values.

For the case where x is odd and $x^3 + 2x^2 = a$, we have $x = 2k + 1$. Therefore, $2k + 3 = n^2$, $k = \frac{n^2 - 3}{2}$. Taking, in turn, $n = 3, 5, 7, \ldots$, we have $k = 3, 11, 23, \ldots$, and $x = 7, 23, 47, \ldots$.

Challenge 1   *Change "odd integer" to "even integer greater than 2,"*
and solve the problem.

Where x is even and $x^3 + 2x^2 = a$, we have $x = 2k$.
Therefore, $2k + 2 = n^2$, $k = \dfrac{n^2 - 2}{2}$. Taking, in turn,
$n = 2, 4, 6, \ldots$, we have $k = 1, 7, 17, \ldots$ and $x = 2,$
$14, 34, \ldots$

Challenge 2   *Change* $x^3 + 2x^2$ *to* $x^3 - 2x^2$, *and then solve the problem.*

Where x is odd and $x^3 - 2x^2 = a$, we have $x = 2k + 1$.
Therefore, $2k - 1 = n^2$, $k = \dfrac{n^2 + 1}{2}$. Taking, in turn,
$n = 1, 3, 5, \ldots$, we have $k = 1, 5, 13, \ldots$ and $x = 3,$
$11, 27, \ldots$.

Challenge 3   *In Challenge* 1 *change* $x^3 + 2x^2$ *to* $x^3 - 2x^2$ *and solve*
the problem.

Where x is even and $x^3 - 2x^2 = a$, we have $x = 2k$.
Therefore, $2k - 2 = n^2$, $k = \dfrac{n^2 + 2}{2}$. Taking, in turn,
$n = 2, 4, 6, \ldots$, we have $k = 3, 9, 19, \ldots$, and $x = 6,$
$18, 38, \ldots$.

8-5 *If* $\dfrac{3x - 5}{x^2 - 1} = \dfrac{A}{x - 1} + \dfrac{B}{x + 1}$ *is true for all permissible values of* x,
find the numerical value of $A + B$.

Multiply the equation by $x^2 - 1$. Then,

$$3x - 5 = A(x + 1) + B(x - 1) = x(A + B) + (A - B).$$
$$\therefore A + B = 3$$

Is there a detectable connection between the sum of the numerators A and B and the original numerator? Let's find out!

$$\frac{Kx + L}{x^2 - 1} = \frac{A}{x - 1} + \frac{B}{x + 1}, \quad Kx + L = x(A + B) + (A - B)$$
$$\therefore A + B = K$$

Verify this conclusion for $\dfrac{2x + 5}{x^2 - 1} = \dfrac{A}{x + 1} + \dfrac{B}{x - 1}$.

What modification in the conclusion that $A + B = K$, if any,
must be made if the problem reads $\dfrac{2x + 5}{x^2 - 4} = \dfrac{A}{x + 2} + \dfrac{B}{x - 2}$?

ANSWER: None

What modification follows if $\dfrac{2x+5}{4x^2-1} = \dfrac{A}{2x+1} + \dfrac{B}{2x-1}$?

ANSWER: $A + B = 1 = \dfrac{K}{2}$

Generalizing further, try $\dfrac{2x+5}{a^2x^2-1} = \dfrac{A}{ax+1} + \dfrac{B}{ax-1}$.

ANSWER: $A + B = \dfrac{2}{a} = \dfrac{K}{a}$

A harder one! $\dfrac{2x+5}{x(x^2-1)} = \dfrac{A}{x} + \dfrac{B}{x+1} + \dfrac{C}{x-1}$

ANSWER: $2x + 5 = x^2(A + B + C) + x(C - B) - A$

$\therefore A + B + C = 0$

An easier one! $\dfrac{x^2-2x-5}{x(x^2-1)} = \dfrac{A}{x} + \dfrac{B}{x-1} + \dfrac{C}{x+1}$

ANSWER: $A + B + C = 1$

Slightly harder! $\dfrac{x^2-2x+5}{2x(x^2-1)} = \dfrac{A}{x} + \dfrac{B}{x-1} + \dfrac{C}{x+1}$

ANSWER: $A + B + C = \dfrac{1}{2}$

Study the last 3 cases for a pattern. Compare the degrees of the polynomials in the numerator and the denominator.

8-6 *For what integral values of* x *and* p *is* $(x^2 - x + 3)(2p + 1)$ *an odd number?*

Since each of the factors is odd for all integers x and p, the product is always an odd number.

8-7 *Express the simplest relation between* a, b, *and* c, *not all equal, if* $a^2 - bc = b^2 - ca = c^2 - ab$.

Since $a^2 - bc = b^2 - ca$, $ca - bc = b^2 - a^2$. Therefore, $c(a - b) = (b + a)(b - a)$. If $a \neq b$ then $c = -(b + a)$, that is, $a + b + c = 0$. What change in procedure do you suggest if $a = b$?

8-8 *Find the two linear factors with integral coefficients of* $P(x, y) = x^2 - 2y^2 - xy - x - y$, *or show that there are no such factors.*

METHOD I: We have $P(x, y) = x^2 - 2y^2 - xy - x - y$. If there are linear factors we may write them as $x + A_1y + B_1$, and $x + A_2y + B_2$, with the numerical values of A_1, A_2, B_1, B_2 to be determined.

Equating the two representations of $P(x, y)$, we have

$$x^2 - 2y^2 - xy - x - y = (x + A_1y + B_1)(x + A_2y + B_2)$$
$$= x^2 + A_1A_2y^2 + (A_1 + A_2)xy + (B_1 + B_2)x$$
$$+ (A_1B_2 + A_2B_1)y + B_1B_2$$

$\therefore -2 = A_1A_2$, $-1 = A_1 + A_2$, $-1 = B_1 + B_2$, $-1 = A_1B_2 + A_2B_1$, $0 = B_1B_2$. Therefore, either $B_1 = 0$ or $B_2 = 0$. (Why not both?) Choose $B_1 = 0$, then $B_2 = -1$, $A_1 = 1$, $A_2 = -2$.

Check to see if $-2 = A_1A_2 = (1)(-2)$.

Therefore, one set of factors is $x + y$ and $x - 2y - 1$. If we choose $B_2 = 0$, we obtain the same set of factors. It follows that this set is unique.

METHOD II: You may be clever enough (or lucky enough) to see that $P(x, y)$ can be written as

$$x^2 - y^2 - y^2 - xy - x - y = (x + y)(x - y) - y(y + x) - (x + y).$$
$$\therefore P(x, y) = (x + y)(x - y - y - 1) = (x + y)(x - 2y - 1).$$

8-9 *Find the sum of the digits of* $(100{,}000 + 10{,}000 + 1000 + 100 + 10 + 1)^2$.

This problem offers a good opportunity to show the advantage of "generalizing" a problem to arrive at a solution. Consider any n-digit number $a_1a_2 \ldots a_n$, $n \le 9$. Then $(a_1 + a_2 + \cdots + a_n)^2$

$$= a_1{}^2 + a_2{}^2 + \cdots + a_n{}^2 + 2a_1a_2 + 2a_1a_3 + \cdots$$
$$+ 2a_2a_3 + \cdots + 2a_{n-1}a_n$$
$$= 2(a_1{}^2 + a_2{}^2 + \cdots + a_n{}^2 + a_1a_2 + a_1a_3 + \cdots$$
$$+ a_{n-1}a_n) - (a_1{}^2 + a_2{}^2 + \cdots + a_n{}^2)$$
$$= 2S_1 - S_2.$$

The number of digits in S_1 is $2 \cdot \dfrac{n}{2}(n + 1)$, and the number of digits in S_2 is n. There are, therefore, $2 \cdot \dfrac{n}{2}(n + 1) - n = n^2$ digits in $(a_1 + a_2 + \cdots + a_n)^2$. Applied to this problem with $n = 6$ and $a_1 = a_2 = a_3 = a_4 = a_5 = a_6 = 1$, the formula yields $6^2 = 36$ as the sum of the digits.

8-10 *How do you invert a fraction, using the operation of addition?*

Let the given fraction be $\frac{a}{b}$, and let x be such that $\frac{a+x}{b+x} = \frac{b}{a}$.
Therefore, $a^2 + ax = b^2 + bx$, $\quad x(a - b) = b^2 - a^2 = (b + a)(b - a)$, $x = -(a + b)$, $a \neq b$.

VERIFICATION: $\dfrac{a - (a + b)}{b - (a + b)} = \dfrac{-b}{-a} = \dfrac{b}{a} = \dfrac{1}{\frac{a}{b}}$

ILLUSTRATION: Let $a = 3$ and $b = 8$; then $x = -11$, and

$$\frac{3 - 11}{8 - 11} = \frac{-8}{-3} = \frac{8}{3} = \frac{1}{\frac{3}{8}}.$$

8-11 *How do you generate the squares of integers from pairs of consecutive integers?*

Possibly the first trial would be the operation of multiplication, but this trial proves disappointing since the product of two consecutive integers $n(n + 1) = n^2 + n \neq m^2$. Why?

Let us try the operation of addition.

METHOD I: We note that $0 + 1 = 1$, $4 + 5 = 9$, $12 + 13 = 25$, and so forth. The form $4 + 5 = 3^2$, $12 + 13 = 5^2$, $24 + 25 = 7^2$, and so forth, suggests the generating function $2mn + (m^2 + n^2) = (m^2 - n^2)^2$, where $m - n = 1$ and, hence, $(m - n)^2 = 1$. For example, for $m = 4$, $n = 3$, $2mn = 24$, $m^2 + n^2 = 25$, and $(m^2 - n^2)^2 = 49$. For $m = 5$, $n = 4$, $2mn = 40$, $m^2 + n^2 = 41$, and $(m^2 - n^2)^2 = 81$.

What is the relation between this generating function and Pythagorean triplets?

METHOD II: We note that $4 = 4 \cdot 1$, $12 = 4 \cdot 3$, $24 = 4 \cdot 6$, $40 = 4 \cdot 10$, and that the second factors 1, 3, 6, 10 are the triangular numbers T_k, with $k = 1, 2, 3, \ldots$. (See Appendix VII.) This suggests the generating function $4T_k + (4T_k + 1) = 4 \cdot \frac{1}{2} k(k + 1) + 4 \cdot \frac{1}{2} k(k + 1) + 1 = (2k + 1)^2$.

For $k = 3$, we have $24 + 25 = 7^2 = 49$,
for $k = 4$, we have $40 + 41 = 9^2 = 81$,
for $k = 5$, we have $60 + 61 = 11^2 = 121$.

8-12 *Is there an integer* N *such that* $N^3 = 9k + 2$, *where* k *is an integer?*

With respect to the divisor 3, N can be expressed as $3m$, $3m + 1$, or $3m - 1$ (alternatively, $3n + 2$) so that N^3 is, respectively, $27m^3$, $27m^3 + 27m^2 + 9m + 1$, or $27m^3 - 27m^2 + 9m - 1$ where m is an integer.

CASE I: $N^3 = 9k + 2 \neq 27m^3$ since, upon dividing both sides of the inequality by 9, the remainder on the left is 2 while the remainder on the right is 0.

CASE II: $N^3 = 9k + 2 \neq 27m^3 + 27m^2 + 9m + 1$ since the respective remainders, upon division by 9, are 2 and 1.

CASE III: $N^3 = 9k + 2 \neq 27m^3 - 27m^2 + 9m - 1$. Why? Therefore, the cube of an integer cannot be expressed in the form $9k + 2$, where k is an integer.

Challenge *Is there an integer* N *such that* $N^3 = 9k + 8$?

$N^3 = 9k + 8 = 9(k + 1) - 1$. Hence, if $N = 3m - 1$, N^3 is of the form $9r - 1$ where $r = 3m^3 - 3m^2 + m$, and, of the form $9k + 8$ where $r = k + 1$.

ILLUSTRATION: $N = 5$ $\therefore$ $N^3 = 5^3 = 9 \cdot 13 + 8 = 125$. Here $N = 3 \cdot 2 - 1$ so that $m = 2$. If $N = -4$, then $N^3 = (-4)^3 = 9(-8) + 8 = -64$. Here $N = 3(-1) - 1$ so that $m = -1$.

8-13 *Let* $S_n = 1^n + 2^n + 3^n + 4^n$, *and let* $S_1 = 1 + 2 + 3 + 4 = 10$. *Show that* S_n *is a multiple of* S_1 *for all natural numbers* n, *except* n = 4k *where* k = 0, 1, 2, ...

We verify directly that $S_1 = 10$, $S_2 = 30 = 3S_1$, $S_3 = 100 = 10S_1$, and $S_4 = 354$, which is not a multiple of S_1. Note also that $S_0 = 4$ is not a multiple of S_1.

We know (Appendix I) that, when a is a positive integer, a, a^5, a^9, . . . , a^{4k+1} all have the same units digit. Therefore, a^{4k+2} TDa^2, a^{4k+3} TDa^3, and a^{4k+4} TDa^4.

It follows that $S_{4k+r} = 1^{4k+r} + 2^{4k+r} + 3^{4k+r} + 4^{4k+r}$ TD $1^r + 2^r + 3^r + 4^r = S_r$, where $r = 1, 2, 3, 4$ (or zero). Since we showed by direct verification that S_r is a multiple of 10 for $r = 1, 2$, and 3, but not for $r = 4$ (or zero), the conclusion is valid.

8-14 *A positive integer* N *is squared to yield* N_1, *and* N_1 *is squared to yield* N_2. *When* N_2 *is multiplied by* N *the result is a seven-digit number ending in* 7. *Find* N.

This problem seems difficult, but it is really quite easy. From the given conditions, the seven-digit number is N^5. We know (see Appendix I) that the units' digit of N^5 is the same as that of N, so that N ends in 7.

Since $7^5 < 10^5 = 100,000$, we conclude that $N \neq 7$. Therefore, N has two digits. (Why not three digits, or more?) We must decide between 17, 27, 37, and so forth.

Since $30^5 = 24, 300,000$, N is less than 30. The choice is now narrowed to 17 or 27. The selection of 17 is based on the fact that the difference between 30^5 and 9,999,999 is very much greater than the difference between 9,999,999 and 20^5.

The answer 17 is unique; it can be verified by actual computation.

8-15 *Let* f = mx + ny, *where* m, n *are fixed positive integers, and* x, y *are positive numbers such that* xy *is a fixed constant. Find the minimum value of* f.

Since m, n are fixed and xy is fixed, $(mx)(ny) = (mn)(xy)$ is a fixed quantity. We now use the theorem that, if the product of two numbers is constant, the minimum sum occurs when the numbers are equal. (See Appendix III.) Therefore, minimum f occurs when $mx = ny$; that is, when $\dfrac{x}{y} = \dfrac{n}{m}$.

ILLUSTRATION 1: Let $m = 5$, $n = 3$, $xy = 60$. Therefore, $x:y = 3:5$, $x: \left(\dfrac{60}{x}\right) = 3:5$.

$$x^2 = 36, \ x = 6, \ y = 10 \quad \therefore f(\min) = 5 \cdot 6 + 3 \cdot 10 = 60$$

ILLUSTRATION 2. Let $m = 5$, $n = 3$, $xy = 100$.

$$f(\min) = 5\sqrt{60} + 3\left(\dfrac{100}{\sqrt{60}}\right) = 10\sqrt{60}$$

9   Diophantine Equations: The Whole Answer

9-1 *A shopkeeper orders* 19 *large and* 3 *small packets of marbles, all alike. When they arrive at the shop, he finds the packets broken open with all the marbles loose in the container. Can you help the shopkeeper make new packets with the proper number of marbles in each, if the total number of marbles is* 224?

Represent the number of marbles in a large packet by L and the number in a small packet by S. Then $19L + 3S = 224$, $S = 74 - 6L + \dfrac{2 - L}{3}$. Since S and L are positive integers, $\dfrac{2 - L}{3}$ must be an integer. If $L = 2$, not a likely value, $\dfrac{2 - L}{3} = 0$; otherwise $\dfrac{2 - L}{3}$ is negative. Let us put $\dfrac{2 - L}{3} = -k$ so that $L = 2 + 3k$. Since $74 - 6L + \dfrac{2 - L}{3} > 0$, $74 > 6(2 + 3k) + k$ so that $k \leq 3$. Also, $S = 74 - 6(2 + 3k) - k = 62 - 19k$. Since $L > S$, $2 + 3k > 62 - 19k$ so that $k > 2$. Since $2 < k \leq 3$, $k = 3$. Therefore, $L = 2 + 3k = 11$ and $S = 62 - 19k = 5$.

The values $L = 11$, $S = 5$ satisfy the conditions of the problem uniquely.

9-2 *Find the integral solutions of* $6x + 15y = 23$.

METHOD I: The left side of the equation is divisible by 3, but the right side is not. Hence, no solutions in integers.

METHOD II: Suppose you overlooked the quick method. A formal procedure yields $x = 3 - 2y + \dfrac{5 - 3y}{6}$. To insure the integral nature of x, $\dfrac{5 - 3y}{6}$ must be an integer. Letting $t = \dfrac{5 - 3y}{6}$, we obtain $y = \dfrac{5}{3} - 2t$. Since t is an integer, y cannot possibly be integral.

Challenge 1 *Solve in positive integers* 13x + 21y = 261.

$x = 20 - y - \dfrac{8y - 1}{13}$. Let $t = \dfrac{8y - 1}{13}$ so that $y = t + \dfrac{5t + 1}{8}$. Let $u = \dfrac{5t + 1}{8}$ so that $t = u + \dfrac{3u - 1}{5}$. Let $v = \dfrac{3u - 1}{5}$ so that $u = v + \dfrac{2v + 1}{3}$. Let $w = \dfrac{2v + 1}{3}$ so that $v = w + \dfrac{w - 1}{2}$.

Since x and y are integers, each of t, u, v, and w is an integer. When $w = 1$, we have in succession $v = 1$, $u = 2$, $t = 3$, $y = 5$, and $x = 12$, obtained by sub-stituting back into the equations given.

The pair $x = 12$, $y = 5$ satisfies the given equation. We now show that the solution is unique. Obviously $21y < 261$ so that $y \le 12$. But $y = \dfrac{13w - 3}{2}$, so $\dfrac{13w - 3}{2} \le 12$, $13w \le 27$, $w \le 2$. But w must be odd. Therefore, the only permissible value is $w = 1$ and, consequently, there is only one value for y.

9-3 *A picnic group transported in n buses (where n > 1 and not prime) to a railroad station, together with 7 persons already waiting at the station, distribute themselves equally in 14 railroad cars. Each bus, nearly filled to its capacity of 52 persons, carried the same number of persons. Assuming that the number of picnickers is the smallest possible for the given conditions, find the number of persons in each railraod car.*

Represent by x the number of persons carried by each bus, and by y the number of persons in each car. Then $nx + 7 = 14y$. Neither x nor n can be an even integer. Since $n \ne 7$, we can try $x = 49$, which is a multiple of 7 and close to 52. With this value for x we have $y = \dfrac{1}{2} + \dfrac{7n}{2}$. Since n must be odd and $n > 1$ and n is neither 3 nor 5 nor 7, we try $n = 9$. Then $y = \dfrac{1}{2} + \dfrac{63}{2} = 32$. Check $14 \times 32 = (49 \times 9) + 7$.

9-4 *Find the number of ways that change can be made of $1.00 with 50 coins (U.S.).*

Let x_1 represent the number of 50¢ pieces,
x_2 represent the number of 25¢ pieces,
x_3 represent the number of 10¢ pieces,
x_4 represent the number of 5¢ pieces,
x_5 represent the number of 1¢ pieces.

First we show that $x_1 = 0$, for if $x_1 = 1$, then the remaining 50¢ of the \$1.00 must be the value of 49 coins, an impossibility. And certainly x_1 cannot exceed 1. The number of pennies $x_5 = 50 - (x_2 + x_3 + x_4)$.

EQUATION 1: $25x_2 + 10x_3 + 5x_4 + 50 - (x_2 + x_3 + x_4) = 100$
EQUATION 2: $5x_2 + 2x_3 + x_4 - \frac{1}{5}(x_2 + x_3 + x_4) = 10$

We now make three observations. In every case the value of any coin used must be a multiple of 5. (See Equation 1.) The value of $25x_2 + 10x_3 + 5x_4$ exceeds 50 so that the value of $50 - (x_2 + x_3 + x_4)$ is less than 50. (See Equation 1.) $x_2 + x_3 + x_4$ is a multiple of 5. (See Equation 2.)

In tabular form the possibilities are shown below.

KIND	NUMBER	VALUE	KIND	NUMBER	VALUE
x_5 (1¢)	45	45	x_5 (1¢)	40	40
x_4 (5¢)	2	10	x_4 (5¢)	8	40
x_3 (10¢)	2	20	x_3 (10¢)	2	20
x_2 (25¢)	1	25		50	100
	50	100			

9-5 *Let* x *be a member of the set* $\{1, 2, 3, 4, 5, 6, 7\}$; y, *a member of the set* $\{8, 9, 10, 11, 12, 13, 14\}$; *and* z, *a member of the set* $\{15, 16, 17, 18, 19, 20, 21\}$. *If a solution of* x + y + z = 33 *is defined as a triplet of integers, one each for* x, y, *and* z *taken from their respective sets, find the number of solutions.*

When $z = 21$, $x + y = 12$; that is, $1 + 11$, $2 + 10$, $3 + 9$, or $4 + 8$, four combinations. Similarly, when $z = 20$, the number of combinations for $x + y$ is 5; when $z = 19$, the number of combinations for $x + y$ is 6; when $z = 18$, the number of combinations for $x + y$ is 7; when $z = 17$, the number of combinations for $x + y$ is 6; when $z = 16$, the number of combinations for $x + y$ is 5; and when $z = 15$; the number of combinations for $x + y$ is 4. In all, there are 37 possibilities.

9-6 R_1, R_2, *and* R_3 *are three rectangles of equal area. The length of* R_1 *is 12 inches more than its width, the length of* R_2 *is 32 inches more than its width, and the length of* R_3 *is 44 inches more than its width. If all dimensions are integers, find them.*

Let $v = k_1u$ and let $w = k_2u$, where u, v, w represent, in inches, the widths of R_1, R_2, and R_3, respectively. Then

$u(u + 12) = k_1u(k_1u + 32) = k_2u(k_2u + 44)$,

$\therefore k_1{}^2u^2 + 32k_1u = k_2{}^2u^2 + 44k_2u$.

$(k_1{}^2 - k_2{}^2)u^2 = (44k_2 - 32k_1)u$, $u \neq 0$

$\therefore u = \dfrac{44k_2 - 32k_1}{k_1{}^2 - k_2{}^2} = \dfrac{6}{k_1 - k_2} - \dfrac{38}{k_1 + k_2}$

Let $k_1 = \dfrac{a}{b}$ and let $k_2 = \dfrac{c}{d}$, $k_1 + k_2 = \dfrac{ad + bc}{bd}$, $k_1 - k_2 = \dfrac{ad - bc}{bd}$. We choose $ad + bc = 38$ and $ad - bc = 2$. (Why not 6?) $ad = 20 = 10 \cdot 2 = 5 \cdot 4$ and $bc = 18 = 9 \cdot 2 = 6 \cdot 3 = 18 \cdot 1$.

Since $0 < k_1 < 1$, $0 < k_2 < 1$, we have

a	d	b	c	k_1	k_2	u	v	w
5	4	6	3	$\frac{5}{6}$	$\frac{3}{4}$	48	40	36, accept,
5	4	18	1	$\frac{5}{18}$	$\frac{1}{4}$	72	20	18, reject,
10	2	18	1	$\frac{10}{18}$	$\frac{1}{2}$	144	80	72, reject.

Therefore, the dimensions are 48 and 60, 40 and 72, 36 and 80.

9-7 *Given* $x^2 = y + a$ *and* $y^2 = x + a$ *where a is a positive integer, find expressions for a that yield integer solutions for x and y.*

$x^2 = y + a$, $y^2 = x + a$, so $x^2 - y^2 = y - x$. Therefore, $(x - y)(x + y) + (x - y) = 0$, $(x - y)(x + y + 1) = 0$. Thus, $x = y$ or $x = -1 - y$. Therefore, either $y^2 - y - a = 0$ giving $y = \dfrac{1 \pm \sqrt{1 + 4a}}{2}$, or $y^2 + y + 1 - a = 0$ giving $y = \dfrac{-1 \pm \sqrt{4a - 3}}{2}$.

For y to be integral, $\sqrt{1 + 4a} = 2n + 1$ and $\sqrt{4a - 3} = 2m - 1$. Thus, $a = n^2 + n = n(n + 1)$, $n = 0, 1, 2, \ldots$, or $a = m^2 - m + 1 = m(m - 1) + 1$, $m = 1, 2, 3, \ldots$.

QUERY: Do these values of a give independent solutions?

9-8 *A merchant has six barrels with capacities of* 15, 16, 18, 19, 20, *and* 31 *gallons. One barrel contains liquid* B, *which he keeps for himself, the other five contain liquid* A, *which he sells to two men so that the quantities sold are in the ratio* 1:2. *If none of the barrels is opened, find the capacity of the barrel containing liquid* B.

The key to the solution of this problem is to think in terms of the excesses over 15 gallons. Thus, the capacities are $15 + 0$, $15 + 1$, $15 + 3$, $15 + 4$, $15 + 5$, $15 + 16$ $(= 30 + 1)$. He cannot keep any barrel with an even excess over 15. (Why?) This leaves only the three possibilities $15 + 1$, $15 + 3$, $15 + 5$, which we try in turn.

We find that the liquid B barrel has capacity 20 gallons, and that the merchant sells 33 gallons $(15 + 18)$ to one man and 66 gallons $(16 + 19 + 31)$ to the other man.

9-9 *Find the number of ordered pairs of integer solutions* (x, y) *of the equation* $\frac{1}{x} + \frac{1}{y} = \frac{1}{p}$, p *a positive integer.*

$\frac{1}{x} + \frac{1}{y} = \frac{1}{p}$, $py + px = xy$, $xy - py - px + p^2 - p^2 = 0$, $(x - p)(y - p) = p^2$.

Let $d_1, d_2, \ldots, d_n$ be the n positive divisors of p^2.

Then $x - p = d_i$, $i = 1, 2, \ldots, n$, $y - p = \frac{p^2}{d_i}$,

and $x - p = -d_i$, $y - p = -\frac{p^2}{d_i}$, yielding $2n$ solutions from which we must exclude the case $x - p = -p$, $y - p = -p$, since these imply $x = 0$, $y = 0$. Therefore, there are $2n - 1$ solutions where n is the number of positive divisors of p^2.

ILLUSTRATION: $\frac{1}{x} + \frac{1}{y} = \frac{1}{6}$ where $6^2 = 36$ has the nine positive divisors 1, 2, 3, 4, 6, 9, 12, 18, 36. The number of pairs of integral solutions is $2 \cdot 9 - 1 = 17$. To find the solutions, solve the 18 pairs of equations $x - p = \pm d_i$, $y - p = \frac{\pm p^2}{d_i}$. The set $x - 6 = -6$, $y - 6 = -6$ yields the unacceptable solution $x = 0$, $y = 0$. The solution pairs are (7, 42), (8, 24), (9, 18), (10, 15), (12, 12), (2, −3), (3, −6), (4, −12), (5, −30), and the inverse pairs. The pair (12, 12) is self-inverse so that the total number of solutions is 17.

9-10 *Express in terms of* A *the number of solutions in positive integers of* x + y + z = A *where* A *is a positive integer greater than 3.*

Consider the following triples (x, y, z): when $x = 1, (1, 1, A - 2)$, $(1, 2, A - 3), \ldots, (1, A - 2, 1)$, are the $A - 2$ possible solutions. When $x = 2, (2, 1, A - 3), (2, 2, A - 4), \ldots, (2, A - 3, 1)$, are the $A - 3$ possible solutions. When $x = 3, (3, 1, A - 4)$, $(3, 2, A - 5), \ldots, (3, A - 4, 1)$, are the $A - 4$ possible solutions. ... When $x = A - 2$, $(A - 2, 1, 1)$, is the only possible solution. Therefore the total number of solutions is $1 + 2 + \cdots + (A - 2) = \frac{1}{2}(A - 2)(A - 1)$. [See Appendix VII.]

ILLUSTRATION: For $A = 6$, the solution triplets are $(1, 1, 4)$; $(1, 4, 1)$; $(4, 1, 1)$; $(1, 2, 3)$; $(1, 3, 2)$; $(2, 1, 3)$; $(2, 3, 1)$; $(3, 1, 2)$; $(3, 2, 1)$; $(2, 2, 2)$; ten triplets. Here we have used only 1, 1, $2k - 2$ and 1, 2, $2k - 3$. The triplets 1, 3, $2k - 4$ and 1, 4, $2k - 5$ yield no new solutions.

9-11 *Solve in integers* ax + by = c *where* a, b, *and* c *are integers*, a < b, *and* 1 ≤ c ≤ b, *with* a *and* b *relatively prime.*

We solve the problem for selected values of c, a, and b.

First we show that if $x = x_0$ and $y = y_0$ is a solution of $ax + by = c$ where a and b are relatively prime, then the equations $x = x_0 - bt$ and $y = y_0 + at$ (where $t = 0, \pm 1, \pm 2, \ldots$) give all the solutions.

PROOF: (1) We have $ax + by = c$ and $ax_0 + by_0 = c$.

(2) Therefore, $ax - ax_0 + by - by_0 = 0$, and so $y - y_0 = \frac{a}{b}(x_0 - x)$.

(3) Since $y - y_0$ is an integer, and since $(a, b) = 1$ (that is, the greatest common divisor of a and b is 1), then b must divide $x_0 - x$; that is, $x_0 - x = bt$ where t is an integer.

(4) Therefore, $x = x_0 - bt$ and $y - y_0 = \frac{a}{b} \cdot bt = at$, and so $y = y_0 + at$.

The proof is completed by showing that if $t = t_1$, then $x_1 = x_0 - bt_1$ and $y_1 = y_0 + at_1$ are solutions.

(5) Substitute x_1 and y_1 into the left side of the equation $ax + by = c$. We have $ax_1 + by_1 = a(x_0 - bt_1) + b(y_0 + at_1) = ax_0 + by_0 - abt_1 + abt_1 = ax_0 + by_0$. Since $ax_0 + by_0 = c$ (from step 1), $ax_1 + by_1 = c$, and so the pair (x_1, y_1) is a solution.

CASE I: $5x - 11y = 1$ (Euclidean Algorithm)

Since $11 = 5 \cdot 2 + 1$ and $5 = 1 \cdot 4 + 1$, we have $1 = 5 - 4 \cdot 1$ (from the second equation) and $1 = 5 - 4(11 - 5 \cdot 2)$, (replacing 1 by $11 - 5 \cdot 2$ from the first equation). Therefore, $1 = 5 \cdot 9 - 11 \cdot 4$. Comparing this equation with the given equation $1 = 5x - 11y$, we have $x = 9$ and $y = 4$ as a solution.

Hence, $x = 9 + 11t$ and $y = 4 + 5t$, $t = 0, \pm 1, \pm 2, \ldots$, comprise all the integer solutions of the given equation.

CASE I: $5x - 11y = 1$ (Permutations)

Rearrange the sequence of integers 1 through 11 so that integer n is associated with $n + 5$ when $n + 5 \leq 11$, and with $n + 5 - 11$ when $n + 5 > 11$, as shown in the two rows below.

Upper Row (natural order) 1 2 3 4 5   6 7 8 9 10 11

Lower Row (permutation)   6 7 8 9 10 11 1 2 3 4   5

Start with 5 (since $a = 5$) in the upper row, go to 10 in the lower row, then to 10 in the upper row, to 4 in the lower row, and so forth, until you reach 1. You will obtain the sequence 5, 10, 4, 9, 3, 8, 2, 7, 1, nine terms in all. Therefore, $x = 9$, and, from the given equation, we find $y = 4$. The general solution is $x = 9 + 11t$ and $y = 4 + 5t$, $t = 0, \pm 1, \pm 2, \ldots$.

CASE II: $5x - 11y = 3$

$\frac{5}{3} x - \frac{11}{3} y = 1$. Let $x = 3X$ and $y = 3Y$, so that $5X - 11Y = 1$. From Case I we have $X = 9 + 11t$ and $Y = 4 + 5t$. Therefore, $x = 27 + 33t$ and $y = 12 + 15t$, $t = 0, \pm 1, \pm 2, \ldots$.

CASE III: $5x - 11y = 25$

Rewrite the equation as $5x - 11(y + 2) = 3$ and let $x = 3X$ and $y + 2 = 3Y$. Thus, $5X - 11Y = 1$. From Case I we have $X = 9 + 11t$ and $Y = 4 + 5t$. Therefore, $x = 27 + 33t$ and $y = 3Y - 2 = 10 + 15t$, $t = 0, \pm 1, \pm 2, \ldots$.

CASE IV: $13x - 9y = 1$

Rewrite the equation as $4x - 9(y - x) = 1$.

Let $x = X$ and let $y - x = Y$ so that $4X - 9Y = 1$.

Since $9 = 4 \cdot 2 + 1$ and $4 = 1 \cdot 4$, then $1 = 4(-2) - 9(-1)$.

Therefore, $X = -2$ and $Y = -1$ is a solution. Therefore, the general solution is $X = -2 + 9t$ and $Y = -1 + 4t$. Therefore, the general solution to the given equation is $x = -2 + 9t$ and $y = -3 + 13t$, $t = 0, \pm 1, \pm 2, \ldots$.

CASE IV: $13x - 9y = 1$ (alternative solution)

Upper Row	1	2	3	4	5	6	7	8	9
Lower Row	5	6	7	8	9	1	2	3	4

The sequence is 4, 8, 3, 7, 2, 6, 1. Using the method shown in the second solution of Case I, associate n with $n + 4$ when $n + 4 \leq 9$, and with $n + 4 - 9$ when $n + 4 > 9$. Since there are seven terms in the sequence, $x = 7 + 9k$, $y = 10 + 13k$, $k = 0, \pm 1, \pm 2, \ldots$.

Reconcile the answers given here with those given under the first solution for Case IV.

---

10 Functions: A Correspondence Course

10-1 *Let* f *be defined as* f$(3n) = $ n $+ $ f$(3n - 3)$ *when* n *is a positive integer greater than* 1, *and* f$(3n) = 1$ *when* n $= 1$. *Find the value of* f(12).

Rewrite f as $f(3n) - f(3n - 3) = n$ and use "telescopic" addition.

$$f(3n) \quad - f(3n - 3) = n$$
$$f(3n - 3) - f(3n - 6) = n - 1$$
$$\vdots \qquad \qquad \vdots \qquad \quad \vdots$$
$$f(6) \quad - \quad f(3) \quad = 2$$

We thus obtain $f(3n) - f(3) = 2 + 3 + \cdots + n$, so that $f(3n) = 1 + 2 + 3 + \cdots + n = \frac{1}{2} n(n + 1)$ since $f(3) = 1$. (See Appendix VII.)

$\therefore f(12) = f(3 \cdot 4) = \frac{1}{2}(4)(5) = 10$

COMMENT: Note that f is a triangular number. (See Appendix VII.)

Challenge *Define* f *to be such that* $f(3n) = n^2 + f(3n - 3)$. *Find* $f(15)$.

Using the method shown in the solution above, obtain
$$f(3n) = \frac{1}{6}n(n + 1)(2n + 1).$$
Therefore, $f(15) = \frac{1}{6}(5)(6)(11) = 55$.

COMMENT: Note that f is the sum of the first n square integers. (See Appendix VII.)

10-2 *If* f *is such that* $f(x) = 1 - f(x - 1)$, *express* $f(x + 1)$ *in terms of* $f(x - 1)$.

Since $f(x) = 1 - f(x - 1)$, $f(x + 1) = 1 - f(x)$.
Therefore, $f(x + 1) = 1 - [1 - f(x - 1)] = f(x - 1)$.

10-3 *Let* f $= ax + b$, g $= cx + d$, x *a real number,* a, b, c, d *real constants.* (a) *Find relations between the coefficients so that* f(g) *is identically equal to* x; *that is,* $f(g) \equiv x$, *and* (b) *show that, when* $f(g) \equiv x$, f(g) *implies* g(f).

(a) $f(g) = a(cx + d) + b = acx + ad + b \equiv x$.
$\therefore ac = 1$, $ad + b = 0$, or $b = -ad$.
(b) $g(f) = c(ax + b) + d = acx + bc + d$.
Since $ac = 1$ and $b = -ad$,
$acx + bc + d = x + c(-ad) + d = x - d + d = x$.
Therefore, $[f(g) \equiv x] \Rightarrow g(f)$. (The symbol $\Rightarrow$ is read "implies.")

10-4 *If* $f(x) = -x^n(x - 1)^n$, *find* $f(x^2) + f(x)f(x + 1)$.

$f(x^2) = -x^{2n}(x^2 - 1)^n$, and $f(x + 1) = -(x + 1)^n(x)^n$.
Therefore, $f(x^2) + f(x)f(x + 1) =$
$-x^{2n}(x^2 - 1)^n - x^n(x - 1)^n[-(x + 1)^n x^n] = 0$

10-5 *The density* d *of a fly population varies directly as the population* N, *and inversely as the volume* V *of usable free space. It is also determined experimentally that the density for a maximum population varies directly as* V. *Express* N *(maximum) in terms of* V.

$d = \frac{k_1 N}{k_2 V} = k_3 \frac{N}{V}$ where k_1, k_2, k_3 are positive constants. Therefore, $N = \frac{Vd}{k_3}$. Since $d(\max) = k_4 V$, $N(\max) = \frac{Vk_4 V}{k_3} = kV^2$ where $k = \frac{k_4}{k_3}$.

10-6 *Given the four elementary symmetric functions* $f_1 = x_1 + x_2 + x_3 + x_4$, $f_2 = x_1x_2 + x_2x_3 + x_3x_4 + x_4x_1$, $f_3 = x_1x_2x_3 + x_2x_3x_4 + x_3x_4x_1 + x_4x_1x_2$, $f_4 = x_1x_2x_3x_4$, *express* $S = \dfrac{1}{x_1} + \dfrac{1}{x_2} + \dfrac{1}{x_3} + \dfrac{1}{x_4}$ *in terms of* f_1, f_2, f_3, f_4.

$$S = \frac{x_2x_3x_4 + x_3x_4x_1 + x_4x_1x_2 + x_1x_2x_3}{x_1x_2x_3x_4} = \frac{f_3}{f_4}$$

10-7 *Let* $f(n) = n(n + 1)$ *where* n *is a natural number. Find the values of* m *and* n *such that* $4f(n) = f(m)$ *where* m *is a natural number.*

Assume $4f(n) = f(m)$, or $4n(n + 1) = m(m + 1)$. Then $4n^2 + 4n = m^2 + m$, $4n^2 + 4n + 1 = m^2 + m + 1$, $(2n + 1)^2 = m^2 + m + 1$. But $m^2 + m + 1$ cannot be the square of an integer. Therefore, there are no natural numbers m and n such that $4f(n) = f(m)$.

INTERPRETATION I: We may say that the product of two successive natural numbers cannot be equal to four times the product of some other pair of successive natural numbers.

INTERPRETATION II: Since

$$(4f(n) = f(m)) \Rightarrow \left(4 \cdot \frac{1}{2} f(n) = \frac{1}{2} f(m)\right)$$
$$\Rightarrow \left(4 \cdot \frac{1}{2} n(n + 1) = \frac{1}{2} m(m + 1)\right),$$

and since $\frac{1}{2} k(k + 1)$ represents the sum of the first k natural numbers (see Appendix VII), we may say that the sum of a given number of natural numbers starting with 1 can never equal four times the sum of some other number of natural numbers starting with 1. Try it!

10-8 *Find the positive real values of* x *such that* $x^{(x^x)} = (x^x)^x$.

Taking logarithms of both sides of the equality to some suitable base, we have $x^x \log x = x \log x^x = x^2 \log x$. Therefore, $(x^x - x^2) \log x = 0$. When $\log x = 0$, $x = 1$, and when $x^x - x^2 = 0$, $x = 2$.

COMMENT: The equation $x^x - x^2 = 0$ is also satisfied by $x = 1$, a value we already have.

10-9 *If* $x = 3 + \cfrac{1}{3 + \cfrac{1}{x}}$ *and* $y = 3 + \cfrac{1}{3 + \cfrac{1}{3 + \cfrac{1}{y}}}$, *find the value of*

$|x - y|$.

The equation $x = 3 + \cfrac{1}{3 + \cfrac{1}{x}}$ is equivalent to $x = 3 + \dfrac{x}{3x + 1}$,

$x \neq 0$, and this equation, in turn, is equivalent to $3x^2 + x = 9x + 3 + x$ which, when simplified, becomes $x^2 - 3x - 1 = 0$. Similarly, the equation for y can be converted to the equivalent equation $y^2 - 3y - 1 = 0$. We may represent either of these equations by $t^2 - 3t - 1 = 0$, using the neutral letter t. One root of this equation is $\dfrac{3 + \sqrt{13}}{2}$. Therefore, each of x and y is the fraction expansion of this root so that $|x - y| = 0$.

VERIFICATION: $x = \dfrac{3 + \sqrt{13}}{2} = \dfrac{6 + \sqrt{13} - 3}{2}$ (replacing 3 by $6 - 3$)

$= 3 + \dfrac{\sqrt{13} - 3}{2} = 3 + \cfrac{1}{\cfrac{2}{\sqrt{13} - 3}}$ $\left(\text{since fraction } \dfrac{a}{b} = \dfrac{1}{\dfrac{b}{a}}\right)$

$= 3 + \cfrac{1}{\dfrac{2(\sqrt{13} + 3)}{4}}$ $\left(\text{since } \dfrac{2}{\sqrt{13} - 3} = \dfrac{2}{\sqrt{13} - 3} \cdot \dfrac{\sqrt{13} + 3}{\sqrt{13} + 3}\right.$

$= \left.\dfrac{2(\sqrt{13} + 3)}{4}\right)$

$= 3 + \cfrac{1}{\dfrac{\sqrt{13} + 3}{2}} = 3 + \dfrac{1}{x}$ $\left(\text{since } x = \dfrac{3 + \sqrt{13}}{2}\right)$. Continuing in

this manner, we obtain $x = 3 + \cfrac{1}{3 + \cfrac{1}{x}}$. In a similar manner, we

develop $y = 3 + \cfrac{1}{3 + \cfrac{1}{3 + \cfrac{1}{y}}}$.

Challenge *Express* $\dfrac{3 + \sqrt{13}}{2}$ *as an (infinite) continued fraction. See Problem* 10-10.

$$\frac{3 + \sqrt{13}}{2} = 3 + \cfrac{1}{3 + \cfrac{1}{3 + \cfrac{1}{3 + \cdots}}}$$

This is obtained by operating in the manner shown in the verification section above.

10-10 *Assuming that the infinite continued fraction* $\cfrac{2}{2+\cfrac{2}{2+\cfrac{2}{2+\cdots}}}$

represents a finite value x, *find* x. *(Technically, we say the infinite continued fraction converges to the value* x.)

$x = \cfrac{2}{2+x}, x^2 + 2x - 2 = 0, x = \sqrt{3} - 1$. We reject the value $x = -\sqrt{3} - 1$, since the original fraction is not negative.

Challenge 1 *Assuming convergence, find*

$$y = -2 + \cfrac{2}{-2 + \cfrac{2}{-2 + \cfrac{2}{-2 + \cdots}}}.$$

METHOD I: $y = -2 + \cfrac{2}{y}, y^2 + 2y - 2 = 0,$

$y = -\sqrt{3} - 1$

METHOD II: $y = -2 + \cfrac{2}{-2 + \cfrac{2}{-2 + \cfrac{2}{-2 + \cdots}}} =$

$$-2 - x = -2 - \sqrt{3} + 1 = -\sqrt{3} - 1$$

NOTE: x has the value stated in Problem 10-10.

10-11 *Find* $\lim\limits_{h \to 0} F$; *that is, the limiting value of* F *as* h *becomes arbitrarily close to zero where* $F = \cfrac{\sqrt{3+h} - \sqrt{3}}{h}$, $h \neq 0$.

$F = \cfrac{\sqrt{3+h} - \sqrt{3}}{h} \cdot \cfrac{\sqrt{3+h} + \sqrt{3}}{\sqrt{3+h} + \sqrt{3}} = \cfrac{h}{h[\sqrt{3+h} + \sqrt{3}]}$

$= \cfrac{1}{\sqrt{3+h} + \sqrt{3}}$. Therefore, $\lim\limits_{h \to 0} F = \cfrac{1}{\sqrt{3} + \sqrt{3}} = \cfrac{1}{2\sqrt{3}}$.

10-12 *Find the limiting value of* $F = \cfrac{x^a - 1}{x - 1}$, $x \neq 1$, *where a is a positive integer, as* x *assumes values arbitrarily close to* 1; *that is, find* $\lim\limits_{x \to 1} F$.

$F = \cfrac{x^a - 1}{x - 1} = \cfrac{(x - 1)(x^{a-1} + x^{a-2} + \cdots + x + 1)}{x - 1}$

$= x^{a-1} + x^{a-2} + \cdots + x + 1, x \neq 1.$

Therefore, $\lim\limits_{x \to 1} F = 1 + 1 + \cdots + 1 + 1 = a.$

ILLUSTRATION: Let $F = \dfrac{x^3 - 1}{x - 1} = \dfrac{(x - 1)(x^2 + x + 1)}{x - 1} = x^2 +$
$x + 1, x \neq 0$. As x takes values arbitrarily close to 1, $x^2 + x + 1$
approaches arbitrarily close to 3.

We invoke the help of a geometric picture, Fig. S10-12.
The graph of F is the parabola $y = x^2 + x + 1$ with the point
common to $x = 1$ deleted. As we approach arbitrarily close to
$x = 1$ along the x-axis, we come arbitrarily close to $y = 3$ along
the y-axis.

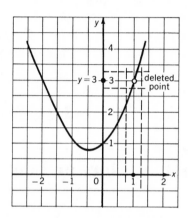

S10-12

10-13 *If* n *is a real number, find* $\lim\limits_{x \to 2} \dfrac{x^n - 2^n}{x - 2}$ *in terms of* n; *that is, the*
limiting value of $\dfrac{x^n - 2^n}{x - 2}$ *as* x *approaches arbitrarily close to 2.*

Let $F = \dfrac{x^n - 2^n}{x - 2} = \dfrac{\left(\dfrac{x}{2}\right)^n - 1}{\left(\dfrac{x}{2} - 1\right)\dfrac{1}{2^{n-1}}}$.   Let $\dfrac{x}{2} = 1 + h$.   Then

$\lim\limits_{x \to 2} F = \lim\limits_{h \to 0} \dfrac{(1 + h)^n - 1}{h} \cdot \dfrac{2^{n-1}}{1}$

$= 2^{n-1} \lim\limits_{h \to 0} \dfrac{\left(1 + nh + \dfrac{n(n - 1)}{1 \cdot 2} h^2 + \cdots\right) - 1}{h}$ (See Appendix VI.)

$= 2^{n-1} \lim\limits_{h \to 0} \left(n + \dfrac{n(n - 1)}{1 \cdot 2} h + \cdots\right) = n \cdot 2^{n-1}$.

10-14 *A function* f *is defined as*

$$f = \begin{cases} 1 \text{ when } x = 1, \\ 2x - 1 + f(x - 1) \text{ when } x \geq 2, x \text{ an integer.} \end{cases}$$

Express f *as the simplest possible polynomial.*

We have $f(2) = 2 \cdot 2 - 1 + 1 = 2^2, f(3) = 2 \cdot 3 - 1 + 2^2 = 3^2$. Let us guess $f = x^2$, and try to prove the result by mathematical induction.

(1) We have $f(1) = 1^2 = 1$.

(2) Assume $f(k) = 2k - 1 + f(k - 1) = k^2$.

$(2k + 1) + 2k - 1 + f(k - 1) = (2k + 1 + k^2)$.

$\therefore 2(k + 1) - 1 + [2k - 1 + f(k - 1)] = (k + 1)^2$;

that is, $2(k + 1) - 1 + f(k) = (k + 1)^2$.

By definition, $f(k + 1) = 2(k + 1) - 1 + f(k)$.

Therefore, $f(k + 1) = (k + 1)^2$, and so the guess is valid.

11 Inequalities, More or Less

11-1 *Let* $P = \left(\frac{1}{a} - 1\right)\left(\frac{1}{b} - 1\right)\left(\frac{1}{c} - 1\right)$ *where* a, b, c *are positive numbers such that* $a + b + c = 1$. *Find the largest integer* N *such that* $P \geq N$.

$P = \left(\frac{1}{a} - 1\right)\left(\frac{1}{b} - 1\right)\left(\frac{1}{c} - 1\right)$

$P = \frac{1}{abc} - \left(\frac{1}{ac} + \frac{1}{bc} + \frac{1}{ab}\right) + \left(\frac{1}{a} + \frac{1}{b} + \frac{1}{c}\right) - 1$

Since $\frac{1}{ac} + \frac{1}{bc} + \frac{1}{ab} = \frac{b + a + c}{abc}$ and $a + b + c = 1$,

$P = \frac{1}{abc} - \frac{1}{abc} + \left(\frac{1}{a} + \frac{1}{b} + \frac{1}{c}\right) - 1$.

Therefore, $P = \frac{1}{a} + \frac{1}{b} + \frac{1}{c} - 1$.

Since $a + b + c = 1$, $(a + b + c)P = P =$

$(a + b + c)\left(\frac{1}{a} + \frac{1}{b} + \frac{1}{c}\right) - 1$.

But $(a + b + c)\left(\frac{1}{a} + \frac{1}{b} + \frac{1}{c}\right) \geq 3^2$. (See Appendix IV.) Therefore, $P \geq 9 - 1 = 8$.

11-2 *Find the pair of least positive integers* x *and* y *such that* 11x − 13y = 1 *and* x + y > 50.

The general solution to $11x - 13y = 1$ is $x = 6 + 13t$, $y = 5 + 11t$ (see Problem 9-11) with $t = 0, \pm1, \pm2, \ldots$.

To satisfy the second condition, we have $x + y = 6 + 13t + 5 + 11t = 11 + 24t$. Since $x + y > 50$, $11 + 24t > 50$, $t \geq 2$. Taking $t = 2$, the least permissible value, we have $x = 6 + 13 \cdot 2 = 32$ and $y = 5 + 11 \cdot 2 = 27$.

11-3 *Is the following set of inequalities consistent? (Consider three inequalities at a time.)*

$$x + y \leq 3, \quad -x - y \geq 0, \quad x \geq -1, \quad -y \leq 2$$

The set of inequalities $x + y \leq 3$, $x \geq -1$, $-y \leq 2$ (the equivalent of $y \geq -2$), determines the triangular region $P_1P_2P_3$. (See Fig. S11-3.) The inequality $-x - y \geq 0$, or its equivalent, $x + y \leq 0$, represents the set of points in the half-plane below the line $x + y = 0$.

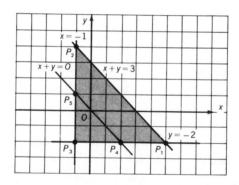

S11-3

The intersection of this half-plane and the triangular region $P_1P_2P_3$ is the triangular region $P_3P_4P_5$.

Since the two triangular regions are not coincident, the set of inequalities is inconsistent.

COMMENT: The inequalities $x + y \leq 3$, $x \geq -1$, and $-y \leq 2$ are consistent, and the set $-x - y \geq 0$, $x \geq -1$, and $-y \leq 2$ is consistent.

11-4 *Find the set of values for* x *such that* $x^3 + 1 > x^2 + x$.

METHOD I: (Algebraic)

$$(x^3 + 1 > x^2 + x) \Rightarrow [(x + 1)(x^2 - x + 1) > x(x + 1)]$$

Therefore, when $x + 1 > 0$, $x^2 - x + 1 > x$.
But $(x^2 - x + 1 > x) \Rightarrow [(x - 1)^2 > 0]$. Therefore, $x \neq 1$.
Since $x + 1 > 0$, $x > -1$. Therefore, $x^3 + 1 > x^2 + x$ for $x > -1$, except for $x = 1$.
Stated otherwise, $x^3 + 1 > x^2 + x$ when $-1 < x < 1$, or when $x > 1$.

METHOD II: (Geometric) In Fig. S11-4 are shown the cubic curve $y = x^3 + 1$ and the parabola $y = x^2 + x$ for $-1 \leq x \leq 1$.

For $x > 1$, $x^3 + 1 > x^2 + x$;
for $x = 1$, $x^3 + 1 = x^2 + x$;
for $-1 < x < 1$, $x^3 + 1 > x^2 + x$;
for $x < -1$, $x^3 + 1 < x^2 + x$.

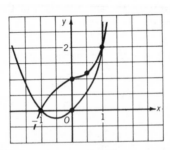

S11-4

11-5 *Consider a triangle whose sides* a, b, c *have integral lengths such that* c < b *and* b ≤ a. *If* a + b + c = 13 *(inches), find all the possible distinct combinations of* a, b, *and* c.

Since $a + b + c = 13$, $b + c = 13 - a$. But $b + c > a$ (the sum of two sides of a non-degenerate triangle is greater than the third side).
Therefore, $13 - a > a$ so that $a \leq 6$. Since $b \leq a$, $b \leq 6$. Therefore, $c \geq 1$.
Also, $c \leq 3$, for if $c \geq 4$, then $a + b \leq 9$; and when $a = b$, a and b each equals, at most, 4. But since $b > c$, this is a contradiction.
When c equals 3, $a + b$ equals 10. Therefore, if $a = b$, each equals 5, and if $a > b$, then $a = 6$ and $b = 4$.

It follows that the possible combinations are

$$a = 6, b = 6, c = 1;$$
$$a = 6, b = 5, c = 2;$$
$$a = 6, b = 4, c = 3;$$
$$a = 5, b = 5, c = 3.$$

In summary, then, $1 \leq c \leq 3$, $4 \leq b \leq 6$, $5 \leq a \leq 6$.

11-6 *A teen-age boy is now* n *times as old as his sister, where* n $> 3\frac{1}{2}$. *In 3 years he will be* n $- 1$ *times as old as she will be then. If the sister's age, in years, is integral, find the present age of the boy.*

Let s represent the sister's age in years; then ns represents the boy's age in years.

$$(n - 1)(s + 3) = ns + 3; \; ns - s + 3n - 3 = ns + 3;$$
$$3n = 6 + s; n = 2 + \frac{s}{3}$$

But $n > 3\frac{1}{2}$, so $2 + \frac{s}{3} > 3\frac{1}{2}$; that is, $s > 4\frac{1}{2}$.

Since the boy is a teen-ager, $13 \leq ns \leq 19$. Substituting for n,

$$13 \leq 2s + \frac{s^2}{3} \leq 19, \text{ and } 39 \leq s^2 + 6s \leq 57.$$

By adding 9 to each side of the inequalities, we have $48 \leq (s + 3)^2 \leq 66$. Therefore, $6 \leq s + 3 \leq 8$, and $3 \leq s \leq 5$.

Since $s > 4\frac{1}{2}$ and $s \leq 5$, $s = 5$, and, since $n = 2 + \frac{s}{3}$, $n = 3\frac{2}{3}$.

Therefore, the present age of the boy, ns, is $18\frac{1}{3}$ years.

11-7 *Express the maximum value of* A *in terms of* n *so that the following inequality holds for any positive integer* n.

$$x^n + x^{n-2} + x^{n-4} + \cdots + \frac{1}{x^{n-4}} + \frac{1}{x^{n-2}} + \frac{1}{x^n} \geq A$$

Since the sum of a positive number and its reciprocal is greater than or equal to 2, $x + \frac{1}{x} \geq 2$ and $x^2 + \frac{1}{x^2} \geq 2$. Therefore,

$x^2 + 1 + \frac{1}{x^2} \geq 3$. Similarly, $x^3 + \frac{1}{x^3} \geq 2$ and $x^3 + x + \frac{1}{x} + \frac{1}{x^3} \geq 4$. Similarly, $x^4 + \frac{1}{x^4} \geq 2$ and $x^4 + x^2 + 1 + \frac{1}{x^2} + \frac{1}{x^4} \geq 5$. Assume that $x^k + x^{k-2} + \cdots + \frac{1}{x^{k-2}} + \frac{1}{x^k} \geq k + 1$, where k is odd. Since $x^{k+2} + \frac{1}{x^{k+2}} \geq 2$, then $x^{k+2} + x^k + x^{k-2} + \cdots + \frac{1}{x^{k-2}} + \frac{1}{x^k} + \frac{1}{x^{k+2}} \geq k + 1 + 2 = k + 3$.   Therefore, $A = n + 1$.

COMMENT: The proof given makes use of Mathematical Induction. (See Appendix VII.)

11-8 *Find the set* $R_1 = \{x | x^2 + (x^2 - 1)^2 \geq |2x(x^2 - 1)|\}$, *and the set* $R_2 = \{x | x^2 + (x^2 - 1)^2 < |2x(x^2 - 1)|\}$.

Let $f = x$ and $g = x^2 - 1$. Since $(f - g)^2 \geq 0$, $f^2 + g^2 \geq 2fg$.

$\therefore x^2 + (x^2 - 1)^2 \geq 2x(x^2 - 1)$

Since $x^2 + (x^2 - 1)^2 > 0$, $x^2 + (x^2 - 1)^2 \geq |2x(x^2 - 1)|$.

Thus, R_1 is the set of real numbers, and R_2 is the empty set.

11-9 *Show that* $F = \frac{1}{2} \cdot \frac{3}{4} \cdot \frac{5}{6} \cdots \frac{99}{100} < \frac{1}{\sqrt{101}}$.

PROOF: Let $G = \frac{2}{3} \cdot \frac{4}{5} \cdot \frac{6}{7} \cdots \frac{100}{101}$.

Since $\frac{1}{2} < \frac{2}{3}$, $\frac{3}{4} < \frac{4}{5}$, $\ldots$, $\frac{k}{k+1} < \frac{k+1}{k+2}$, $\ldots$, $F < G$.

$\therefore F^2 < FG = \frac{1}{2} \cdot \frac{2}{3} \cdot \frac{3}{4} \cdots \frac{99}{100} \cdot \frac{100}{101} = \frac{1}{101}$, and $F < \frac{1}{\sqrt{101}}$.

Challenge   *Show that* $P = \frac{2}{3} \cdot \frac{4}{5} \cdot \frac{6}{7} \cdots \frac{100}{101} > \frac{\sqrt{101}}{100}$.

Since $\frac{1}{2} < \frac{2}{3}$, $\frac{3}{4} < \frac{4}{5}$, $\ldots$, $F < P$.

But $FP = \frac{1}{101}$, $\therefore P > \frac{1}{101} \div \frac{1}{\sqrt{101}}$ and $P > \frac{\sqrt{101}}{101}$.

11-10 *Which is larger* $\sqrt[9]{9!}$ *or* $\sqrt[10]{10!}$ *? Be careful!*

METHOD I: We prove that the positive geometric mean (see Appendix IV) $G_{n+1} = \sqrt[n+1]{(n+1)!}$ is greater than $G_n = \sqrt[n]{n!}$, and, hence, $\sqrt[10]{10!} > \sqrt[9]{9!}$.

Since $\sqrt[n+1]{n+1} > \sqrt[n+1]{n}$, $\sqrt[n+1]{n!(n+1)} > \sqrt[n+1]{n!n}$ (multiplying by $\sqrt[n+1]{n!}$); that is, $\sqrt[n+1]{(n+1)!} > \sqrt[n+1]{n!n}$.

Since $n^n > n!$ for $n > 1$, $n > \sqrt[n]{n!}$.

$\therefore n!n > n!(n!)^{\frac{1}{n}} = (n!)^{\frac{n+1}{n}}$, and $\sqrt[n+1]{n!n} > \sqrt[n]{n!}$.

Thus $\sqrt[n+1]{(n+1)!} > \sqrt[n+1]{n!n} > \sqrt[n]{n!}$.

METHOD II: Raise both expressions to the 90th power.

$$(\sqrt[9]{9!})^{90} \overset{?}{\lessgtr} (\sqrt[10]{10!})^{90}$$
$$(9!)^{10} \overset{?}{\lessgtr} (10!)^{9}$$
$$(9!)^{9}(9!) \overset{?}{\lessgtr} (9!)^{9}(10)^{9}$$
$$9! \overset{?}{\lessgtr} 10^{9}$$

This is obviously true, so we proceed to conclude that

$$\sqrt[9]{9!} < \sqrt[10]{10!}.$$

11-11 *If* x *is positive, how large must* x *be so that* $\sqrt{x^2 + x} - x$ *shall differ from* $\frac{1}{2}$ *by less than* 0.02?

$\left| \sqrt{x^2 + x} - x - \frac{1}{2} \right| < \epsilon$, so we replace .02 by ϵ and, hence, solve a more general problem.

$\frac{1}{2} - \epsilon < \sqrt{x^2 + x} - x < \frac{1}{2} + \epsilon$.

But $\sqrt{x^2 + x} - x = \dfrac{x}{\sqrt{x^2 + x} + x} = \dfrac{1}{\sqrt{1 + \frac{1}{x}} + 1}$.

$\therefore \dfrac{1}{\sqrt{1 + \frac{1}{x}} + 1} > \dfrac{1}{2} - \epsilon = \dfrac{1 - 2\epsilon}{2}$, and $\sqrt{1 + \frac{1}{x}} + 1 < \dfrac{2}{1 - 2\epsilon}$;

$\sqrt{1 + \frac{1}{x}} < \dfrac{1 + 2\epsilon}{1 - 2\epsilon}$; $1 + \dfrac{1}{x} < \dfrac{(1 + 2\epsilon)^2}{(1 - 2\epsilon)^2}$;

$\dfrac{1}{x} < \dfrac{8\epsilon}{(1 - 2\epsilon)^2}$ $\therefore x > \dfrac{(1 - 2\epsilon)^2}{8\epsilon}$.

For $\epsilon = .02$, $x > 5.76$.

11-12 *Find a rational approximation* $\dfrac{m}{n}$ *to* $\sqrt{2}$ *such that* $-\dfrac{1}{8n} <$ $\sqrt{2} - \dfrac{m}{n} < \dfrac{1}{8n}$ *where* $n \leq 8$.

Since $-\frac{1}{8n} < \sqrt{2} - \frac{m}{n} < \frac{1}{8n}$, $-\frac{1}{8} < n\sqrt{2} - m < \frac{1}{8}$. Since $\frac{1}{8} = .125$, we list below, for $n \leq 8$, the approximate values (three decimal places) of $n\sqrt{2} - m$ such that $n\sqrt{2} - m < 1$.

$$\begin{array}{ll} \sqrt{2} - 1 = .414 & 5\sqrt{2} - 7 = .070 \\ 2\sqrt{2} - 2 = .828 & 6\sqrt{2} - 8 = .484 \\ 3\sqrt{2} - 4 = .242 & 7\sqrt{2} - 9 = .898 \\ 4\sqrt{2} - 5 = .656 & 8\sqrt{2} - 11 = .312 \end{array}$$

From the table we find $-\frac{1}{8} < 5\sqrt{2} - 7 < \frac{1}{8}$. Therefore, $-\frac{1}{8 \cdot 5} < \sqrt{2} - \frac{7}{5} < \frac{1}{8 \cdot 5}$ so that the required $\frac{m}{n}$ is $\frac{7}{5}$.

QUERY: Can you show that the answer is unique?

11-13 *Find the least value of* $(a_1 + a_2 + a_3 + a_4)\left(\frac{1}{a_1} + \frac{1}{a_2} + \frac{1}{a_3} + \frac{1}{a_4}\right)$ *where each* a_i, $i = 1, 2, 3, 4$, *is positive.*

We prove more generally that

$$(a_1 + a_2 + \cdots + a_n)\left(\frac{1}{a_1} + \frac{1}{a_2} + \cdots + \frac{1}{a_n}\right) \geq n^2$$

where each a_i, $i = 1, 2, \ldots, n$, is positive so that the answer to the given problem is $4^2 = 16$.

By definition, the harmonic mean (H.M., see Appendix IV) of positive numbers is H.M. $= \left(\dfrac{a_1^{-1} + a_2^{-1} + \cdots + a_n^{-1}}{n}\right)^{-1} = \dfrac{n}{\dfrac{1}{a_1} + \dfrac{1}{a_2} + \cdots + \dfrac{1}{a_n}}$. Since the H.M. $\leq$ A.M. (arithmetic mean; see Appendix IV), then $\dfrac{n}{\dfrac{1}{a_1} + \dfrac{1}{a_2} + \cdots + \dfrac{1}{a_n}} \leq \dfrac{a_1 + a_2 + \cdots + a_n}{n}$.

Therefore, $(a_1 + a_2 + \cdots + a_n)\left(\dfrac{1}{a_1} + \dfrac{1}{a_2} + \cdots + \dfrac{1}{a_n}\right) \geq n^2$.

12 Number Theory: Divide and Conquer

12-1 *Let* $N_1 = .888\ldots$, *written in base* 9, *and let* $N_2 = .888\ldots$, *written in base* 10. *Find the value of* $N_1 - N_2$ *in base* 9.

METHOD I: $N_1 = \dfrac{8}{9} + \dfrac{8}{9^2} + \cdots = 1, N_2 = \dfrac{8}{10} + \dfrac{8}{10^2} + \cdots = \dfrac{8}{9}.$
$\therefore N_1 - N_2 = \dfrac{1}{9}$ which is .1 in base 9.

METHOD II: Since $N_2 = .888\ldots$ in base 10, then $10N_2 = 8.888\ldots$.
By subtraction, $9N_2 = 8$ so that $N_2 = \dfrac{8}{9}.$
Similarly, since $N_1 = .888\ldots$ in base 9, $9N_1 = 8.888\ldots$, and, again by subtraction, $8N_1 = 8$ so that $N_1 = 1$
$\therefore N_1 - N_2 = \dfrac{1}{9}.$

12-2 *Solve* $x^2 - 2x + 2 \equiv 0$ (*mod* 5).

Since $x^2 - 2x + 2 \equiv 0$ (mod 5), $x^2 - 2x + 1 \equiv -1$ (mod 5). We may now replace -1 by $4 - 5$ and, then, reduce the coefficients by multiples of 5. We have $x^2 - 2x + 1 = 4 - 5$ (mod 5), $x^2 - 2x + 1 \equiv 4 - 0$ (mod 5), and, so, $x^2 - 2x + 1 \equiv 4$ (mod 5). Therefore, $x - 1 \equiv 2$ (mod 5) or $x - 1 \equiv -2$ (mod 5).

From $x - 1 \equiv 2$ (mod 5) we obtain $x \equiv 3$ (mod 5); that is, $x = 3 + 5k$.

From $x - 1 \equiv -2$ (mod 5) we obtain $x \equiv -1$ (mod 5). This may be modified to $x \equiv 4 - 5$ (mod 5) and, consequently, $x \equiv 4$ (mod 5); that is, $x = 4 + 5k$.

In either expression for x, k may have the values $0, \pm 1, \pm 2, \ldots$.

12-3 *Find the positive digit divisors, other than* 1, *of* $N = 664{,}512$ *written in base* 9.

As a general principle, it can be said that $N = a_0x^n + a_1x^{n-1} + \cdots + a_{n-1}x + a_n$ is divisible by $x - 1$ if $a_0 + a_1 + \cdots + a_n$ is divisible by $x - 1$, (see proof below), where x, a positive integer, represents the base designated.

Since $6 + 6 + 4 + 5 + 1 + 2 = 24 = 8 \cdot 3$, N is divisible by 8 and, hence, also by 2 and by 4.

PROOF: $N = a_0[(x - 1) + 1]^n + a_1[(x - 1) + 1]^{n-1} + \cdots + a_{n-1}[(x - 1) + 1] + a_n$

We may write $a_0[(x - 1) + 1]^n = a_0 A_0(x - 1) + a_0$

where A_0 is a polynomial in x of degree $n - 1$.

Similarly, $a_1[(x - 1) + 1]^{n-1} = a_1 A_1(x - 1) + a_1$, where A_1 is a polynomial in x of degree $n - 2$. And so forth until we reach $a_{n-2}[(x - 1) + 1]^2 = a_{n-2}A_{n-2}(x - 1) + a_{n-2}$

with A_{n-2} of degree 1. Finally, we have

$$a_{n-1}[(x - 1) + 1] = a_{n-1}(x - 1) + a_{n-1}.$$

N, therefore, is the sum of a multiple of $(x - 1)$

and $a_0 + a_1 + \cdots + a_{n-1} + a_n$. If, then,

$a_0 + a_1 + \cdots + a_{n-1} + a_n$ is divisible by $x - 1$,

then so is N.

NOTE: There may be other digit divisors in special cases. For example, $M = 664{,}422$ in base 9 has an additional divisor of 5. Show that an additional divisor of 5 occurs whenever the base 9 number is divisible by 11 in base 9. (See Appendix V.)

12-4 *Find all the positive integral values of* n *for which* $n^4 + 4$ *is a prime number.*

$n^4 + 4 = n^4 + 4n^2 + 4 - 4n^2 = (n^2 + 2n + 2)(n^2 - 2n + 2)$

For $n = 1$, $n^4 + 4 = 5 \cdot 1$, a prime.

For $n > 1$, $n^2 + 4$ is composite since it has the two factors $n^2 + 2n + 2$ and $n^2 - 2n + 2$, each greater than 1.

Hence, $n^4 + 4$ is prime only for $n = 1$.

12-5 *Let* $B_a = x^a - 1$ *and let* $B_b = x^b - 1$ *with* a, b *positive integers. If* $B_y = x^y - 1$ *is the binomial of highest degree dividing each of* B_a *and* B_b, *how is* y *related to* a *and* b?

Obviously, y divides both a and b because if $a = ya_1$ and a_1 is an integer, then $x^a - 1 = x^{ya_1} - 1 = (x^y)^{a_1} - 1$, which is divisible by $x^y - 1$. A similar argument follows for $b = yb_1$. Thus, the maximum value of y is the greatest common factor of a and b, or $y = (a, b)$.

ILLUSTRATION: Let $a = 9$ and $b = 6$. Then $y = (9, 6) = 3$.

VERIFICATION: $B_a = x^9 - 1 = (x^3)^3 - 1$

$$= (x^3 - 1)(x^6 + x^3 + 1)$$

$$B_b = x^6 - 1 = (x^3)^2 - 1$$

$$= (x^3 - 1)(x^3 + 1)$$

B_a and B_b are each divisible by $B_y = x^3 - 1$.

12-6 *If* $f(x) = x^4 + 3x^3 + 9x^2 + 12x + 20$, *and* $g(x) = x^4 + 3x^3 + 4x^2 - 3x - 5$, *find the functions* $a(x)$, $b(x)$ *of smallest degree such that* $a(x)f(x) + b(x)g(x) = 0$.

Since $f(x) = x^4 + 3x^3 + 9x^2 + 12x + 20$

$$= x^4 + 3x^3 + 5x^2 + 4x^2 + 12x + 20$$

$$= x^2(x^2 + 3x + 5) + 4(x^2 + 3x + 5),$$

and since $g(x) = x^4 + 3x^3 + 4x^2 - 3x - 5$

$$= x^4 + 3x^3 + 5x^2 - (x^2 + 3x + 5)$$

$$= x^2(x^2 + 3x + 5) - (x^2 + 3x + 5),$$

$f(x) = (x^2 + 4)(x^2 + 3x + 5)$,

and $g(x) = (x^2 - 1)(x^2 + 3x + 5)$.

$\therefore b(x) = x^2 + 4$, and $a(x) = -(x^2 - 1) = 1 - x^2$;

or $b(x) = -(x^2 + 4)$, and $a(x) = x^2 - 1$,

since $(1 - x^2)(x^2 + 4)(x^2 + 3x + 5)$

$$+ (x^2 + 4)(x^2 - 1)(x^2 + 3x + 5) = 0,$$

and $(x^2 - 1)(x^2 + 4)(x^2 + 3x + 5)$

$$- (x^2 + 4)(x^2 - 1)(x^2 + 3x + 5) = 0.$$

COMMENT 1: Obviously, if $f(x)$ and $g(x)$ are relatively prime, then $a(x) = -g(x)$ and $b(x) = f(x)$, or $a(x) = g(x)$ and $b(x) = -f(x)$.

COMMENT 2: In general, if $f(x) = D_1(x)Q(x)$ and $g(x) = D_2(x)Q(x)$, then $a(x) = D_2(x)$ and $b(x) = -D_1(x)$, or the respective negatives.

12-7 *Find the smallest positive integral value of* k *such that* $kt + 1$ *is a triangular number when* t *is a triangular number. (See Appendix VII.)*

Since a triangular number t is of the form $\frac{1}{2}n(n + 1)$ where n is a natural number, and since we require that $kt_1 + 1 = t_2$, we have $k\left[\frac{1}{2}r(r+1)\right] + 1 = \frac{1}{2}s(s+1)$. Therefore, $kr^2 + kr + 2 = s(s+1) = (A+1)(A+2)$ where s is replaced by $A+1$. Therefore, $kr^2 + kr + 2 = A^2 + 3A + 2$ so that $kr^2 = A^2$ and $kr = 3A$. Therefore, $r = \frac{A}{3} = \frac{s-1}{3}$.

Since the smallest value of k is required, choose $r = 1$. Then $s = 4$ and $k = 9$.

ILLUSTRATION: Let $t = \frac{1}{2}(5)(6) = 15$; then $9t + 1 = 136 = \frac{1}{2}(16)(17)$.

VERIFICATION: Since $m = 3n + 1$, we have $k \cdot \frac{1}{2}n(n+1) = \frac{1}{2}(3n+1)(3n+2) = \frac{1}{2}(9n^2 + 9n + 2) = 9 \cdot \frac{1}{2}n(n+1) + 1$. Thus, $k = 9$.

12-8 *Express the decimal .3 in base 7.*

We have $\frac{3}{10} = \frac{a_1}{7} + \frac{a_2}{7^2} + \frac{a_3}{7^3} + \cdots$ where a_1, a_2, a_3, $\ldots$ are to be determined. To find a_1, multiply through by 7, thus obtaining

$$\frac{3 \cdot 7}{10} = \frac{21}{10} = 2 + \frac{1}{10} = a_1 + \frac{a_2}{7} + \frac{a_3}{7^2} + \cdots, \text{ so that } a_1 = 2.$$

To find a_2, multiply through again by 7, thus obtaining

$$\frac{1 \cdot 7}{10} = \frac{7}{10} = 0 + \frac{7}{10} = a_2 + \frac{a_3}{7} + \cdots, \text{ so that } a_2 = 0.$$

Continuing in this manner, we have

$$\frac{7 \cdot 7}{10} = \frac{49}{10} = 4 + \frac{9}{10} = a_3 + \frac{a_4}{7} + \cdots, \text{ so that } a_3 = 4,$$

$$\frac{9 \cdot 7}{10} = \frac{63}{10} = 6 + \frac{3}{10} = a_4 + \frac{a_5}{7} + \cdots, \text{ so that } a_4 = 6.$$

Thereafter, the digits repeat in cycles of 2046.

Therefore, .3 (base 10) = .20462046 ... (base 7).

12-9 *The following excerpt comes from Lewis Carroll's* Alice's Adventures in Wonderland.

"Let me see: four times five is twelve, and four times six is thirteen, and four times seven is—oh dear! I shall never get to twenty at that rate!"

Do you agree or disagree with the author?

Here is one interpretation of the excerpt:

4 × 5 (base 10) = 12 (base 18), 4 × 6 (base 10) = 13 (base 21), 4 × 7 (base 10) = 14 (base 24), . . . , 4 × 12 (base 10) = 19 (base 39)

where the successive bases are in arithmetic progression. The next term, if we continue in this vein, should be 4 × 13 (base 10) = 20 (base 42). However, 20 (base 42) = 84, not 52.

If we write 4 × 13 = 1u (base 42), and allow u = 10, we can satisfy the requirement, but if u is limited to the set $\{0, 1, 2, \ldots, 9\}$, we cannot satisfy the requirement.

12-10 *Show that, if* $a^2 + b^2 = c^2$, a, b, c *integers, then* P = abc *is divisible by* 60 = $3 \cdot 4 \cdot 5$.

Let $a = m^2 - n^2$, $b = 2mn$, $c = m^2 + n^2$. Then

$$P = 2mn(m - n)(m + n)(m^2 + n^2).$$

CASE I: If either m or n is even, then $4P$, where $4P$ means that 4 divides P exactly. If both m, n are odd, then $m - n$ is even and, therefore, $4|P$.

CASE II: If either m or n is of the form $3k$, then $3|P$.
If $m = 3k + 1$ and $n = 3L + 1$, then $m - n = 3r$ so that $3|P$.
If $m = 3k + 1$ and $n = 3L - 1$, then $m + n = 3s$ so that $3|P$.
Similarly for other combinations.

CASE III: If either m or n is of the form $5k$, then $5|P$.
If $m = 5k + 1$ and $n = 5L + 1$, then $m - n = 5r$ so that $5|P$.
If $m = 5k + 1$ and $n = 5L + 2$, then $m^2 + n^2 = 5t$ so that $5|P$.
For other combinations, proceed in similar fashion.

These cases are independent and, hence, the results may be superimposed so that P is divisible by 3, by 4, and by 5, and, hence, by $3 \cdot 4 \cdot 5 = 60$. For primes beyond 5, the reasoning fails. Therefore, the largest integer divisor is 60. Alternatively, since the greatest common factor of $3 \cdot 4 \cdot 5$ and $5 \cdot 12 \cdot 13$ is 60, the largest integer divisor is 60.

12-11 *Find the integer values of* x *between* −10 *and* +15 *such that* P = $3x^3 + 7x^2$ *is the square of an integer.*

Since $P = 3x^3 + 7x^2 = x^2(3x + 7) = N^2$, either $x = 0$, or $3x + 7$ is the square of an integer. We therefore set $3x + 7 = K^2 = 3(x + 2) + 1$. Since the right side of this last equality leaves a remainder of 1 when divided by 3, the same holds for K^2. Therefore, $K^2 = 3m + 1$ and, in consequence, $K = 3m \pm 1$. Put another way, since $3x + 7 \equiv 1 \pmod 3$, $K^2 \equiv 1 \pmod 3$.

For $m = 0$, $K = \pm 1$ and $x = -2$.

For $m = 1$, $K = 2$ or 4 and $x = -1$ or 3.

For $m = 2$, $K = 5$ or 7 and $x = 6$ or 14.

Therefore, the required set is $\{-2, -1, 0, 3, 6, 14\}$.

12-12 *Find the geometric mean of the positive divisors of the natural number* n. *(See Appendix IV.)*

Let the divisors of n be $d_0 (= 1), d_1, d_2, \ldots, d_{k-1}, d_k (= n)$ where $d_{k-1} = \dfrac{n}{d_1}$, $d_{k-2} = \dfrac{n}{d_2}$, and so forth. Therefore, when k is even, G.M. $= \sqrt[k]{d_0 d_1 \cdots d_{k-1} d_k} = \sqrt[k]{n \cdot n \cdots n(k/2 \text{ factors})}$, since $n = d_0 d_k = d_1 d_{k-1} = \cdots$. When k is odd there will be $\dfrac{k-1}{2}$ factors n and one factor $\sqrt{n}$. In either case, we have $k\sqrt{n^{\frac{k}{2}}} = n^{\frac{1}{2}} = \sqrt{n}$.

ILLUSTRATION 1: Find the G.M. of the positive divisors of 72.

$$\text{G.M.} = \sqrt[12]{1 \cdot 2 \cdot 3 \cdot 4 \cdot 6 \cdot 8 \cdot 9 \cdot 12 \cdot 18 \cdot 24 \cdot 36 \cdot 72}$$
$$= \sqrt[12]{72^6} = \sqrt{72}$$

ILLUSTRATION 2: Find the G.M. of the positive divisors of 16.

$$\text{G.M.} = \sqrt[5]{1 \cdot 2 \cdot 4 \cdot 8 \cdot 16} = \sqrt[5]{16^2 \cdot 4} = \sqrt[5]{4^5} = 4 = \sqrt{16}$$

12-13 *Show that if* $P = 1 \cdot 2 \cdot 3 \cdot \ldots \cdot n$ *and* $S = 1 + 2 + 3 + \cdots + n$, n *a natural number, then* S *exactly divides* P *if* n *is odd.*

PROOF: If n is odd we may represent it as $2k + 1$. Then

$$\frac{P}{S} = \frac{1 \cdot 2 \cdot 3 \cdot \ldots (2k + 1)}{1 + 2 + 3 + \cdots + (2k + 1)} = \frac{(2k + 1)!}{\frac{1}{2}(2k + 1)(2k + 2)}$$
$$= \frac{(2k + 1)!}{(2k + 1)(k + 1)} \qquad \text{(See Appendix VII.)}$$

But both $(2k + 1)$ and $(k + 1)$ are factors of $(2k + 1)!$. Therefore, S divides P.

If n is even we may represent it as $2k$. Then $\dfrac{P}{S} = \dfrac{(2k)!}{\frac{1}{2}(2k)(2k+1)}$,

which may or may not be an integer since $2k + 1$ is not necessarily a factor of $(2k)!$.

ILLUSTRATION 1: Let $n = 21$.

Then $P = 21!$ and $S = \frac{1}{2}(21)(22) = 11 \cdot 21$.

Since 21! contains each of the factors 11 and 21, S divides P.

ILLUSTRATION 2: Let $n = 8$. Then $P = 8!$ and $S = \frac{1}{2}(8)(9) = 36$, and 8! is divisible by 36.

ILLUSTRATION 3: Let $n = 6$. Then $P = 6!$ and $S = 21$, and 6! is not divisible by 21.

12-14 *By shifting the initial digit 6 of the positive integer* N *to the end, we obtain a number equal to* $\frac{1}{4}$ N. *Find the smallest possible value of* N *that satisfies the conditions.*

Let the digit representation of N be $6a_2a_3 \ldots a_n$ so that $a_2a_3 \ldots a_n6 = \frac{1}{4}(6a_2a_3 \ldots a_n)$. When each side of this equation is multiplied by 4, the terminal digit on the right is a_n while the terminal digit on the left is 4. Thus, $a_n = 4$. Then, the digit preceding 6, on the left side, is 8 since $4 \times 4 + 2 = 18$, so that the corresponding digit on the right $a_{n-1} = 8$. Continuing in this manner, we have $4 \times 8 + 1 = 33$ so that $a_{n-2} = 3$, $4 \times 3 + 3 = 15$ so that $a_{n-3} = 5$, $4 \times 5 + 1 = 21$ so that $a_{n-4} = 1$, and, finally, $4 \times 1 + 2 = 6$ so that $a_{n-5} = 6$. Therefore, $N = 615,384$.

VERIFICATION: $\frac{1}{4}(615,384) = 153,846$

Of course larger values of N are obtainable by repeating the basic block of integers which, in this case, are 6, 1, 5, 3, 8, 4. Thus, for example, we have

$$N_2 = 615,384,615,384 \quad \text{or} \quad N_3 = 615,384,615,384,615,384,$$

and so forth, each satisfying the conditions of the problem, since, for example, $\frac{1}{4}(615,384) = 153,846$ and $\frac{1}{4}(615,384,000,000) = 153,846,000,000$, and, hence, $\frac{1}{4}(615,384,000,000) + \frac{1}{4}(615,384) = 153,846,000,000 + 153,846$; that is,

$$153,846,153,846 = \frac{1}{4}(615,384,615,384).$$

Show that the general form for N is $615,384(10^{6m} + 10^{6(m-1)} + \cdots + 10^6 + 1)$, where $m = 0, 1, 2, 3 \ldots$.

12-15 *Find the two-digit number* N (*base* 10) *such that when it is divided by* 4 *the remainder is zero, and such that all of its positive integral powers end in the same two digits as the number.*

Set $N = 10a + b$. Since $10a + b = 4m$, b is even. The only two even digits whose square has the same terminating digit as the digit itself are 0 and 6. Hence. $b = 0$ or 6.

The case $b = 0$ leads to $a = 0$ so that $N = 00$, a trivial case, for, if $a \neq 0$, N will terminate in 0 while its square will terminate in 00.

$N = 10a + 6 = 4m$, $5a + 3 = 2m$; $\therefore a = 1, 3, 5, 7,$ or 9.

But $N^2 = (10a + 6)^2 = 100a^2 + 120a + 36 = 100a^2 + 100d + 10e + 36$, where we set $120a = 100d + 10e$. Since the last two digits of N^2 are the same as those of N, $10e + 36 = 10a + 6$, $a = e + 3$ so that $a \geq 3$. Also, $120a = 100d + 10(a - 3)$, $11a = 10d - 3$, $11a \leq 87$, $a \leq 7$.

Try $a = 3$, $36^2 = 1296$ (reject). Try $a = 5$, $56^2 = 3136$ (reject). Try $a = 7$, $76 = 5776$ (accept). $\therefore N = 76$.

12-16 *Find a base* b *such that the number* 321_b (*written in base* b) *is the square of an integer written in base* 10.

Since $3b^2 + 2b + 1 = N^2$, $b = \dfrac{-2 + \sqrt{4 - 12 + 12N^2}}{6}$ (quadratic formula). Since $b \geq 4$ and integral (Why?), the expression $-2 + \sqrt{12N^2 - 8} \geq 6k$ with $k = 4, 5, \ldots$. The values $k = 4$ and $k = 5$ yield non-integral values for N. For $k = 6$, $-2 + \sqrt{12N^2 - 8} = 36$ and $N^2 = 121$.

VERIFICATION: $321_6 = 121_{10} = 11^2$

12-17 *If* $\dfrac{(a - b)(c - d)}{(b - c)(d - a)} = -\dfrac{5}{3}$, *find* $\dfrac{(a - c)(b - d)}{(a - b)(c - d)}$.

Let $\dfrac{(a - b)(c - d)}{(b - c)(d - a)} = f$; then $\dfrac{(b - c)(d - a)}{(a - b)(c - d)} = \dfrac{1}{f}$,

and $1 - \dfrac{(b - c)(d - a)}{(a - b)(c - d)} = 1 - \dfrac{1}{f} = \dfrac{(a - b)(c - d) - (b - c)(d - a)}{(a - b)(c - d)}$

$= \dfrac{ac - bc - ad + bd - bd + ab + cd - ac}{(a - b)(c - d)}$

$= \dfrac{(a - c)(b - d)}{(a - b)(c - d)} = 1 - \dfrac{1}{f} = 1 + \dfrac{3}{5} = \dfrac{8}{5}$

12-18 *Solve* $x(x + 1)(x + 2)(x + 3) + 1 = y^2$ *for integer values of* x *and* y.

Let $P = x(x + 1)(x + 2)(x + 3)$. Since the product of four consecutive integers is divisible by 24, we may write $P = 24m$ where $m = 0, 1, 2, \ldots$. Since $P + 1 = 24m + 1$ is the square of an integer for selected values of m, we set

$$x(x + 1)(x + 2)(x + 3) + 1$$
$$= x^4 + 6x^3 + 11x^2 + 6x + 1 = (x^2 + ax + 1)^2.$$

By comparing the coefficients of like powers of x on both sides of this identity, we find $a = 3$.

Therefore, $y^2 = (x^2 + 3x + 1)^2$ so that $y = x^2 + 3x + 1$ or $y = -(x^2 + 3x + 1)$, and, hence, there are infinitely many solutions in integers since we may assign to x any arbitrary integer value.

12-19 *Factor* $x^4 - 6x^3 + 9x^2 + 100$ *into quadratic factors with integral coefficients.*

METHOD I: Let $P = x^4 - 6x^3 + 9x^2 + 100 = x^2(x^2 - 6x + 9) + 100 = x^2(x - 3)^2 + 100$. If $x^2 + ax + b$ is a factor of P, then a value of x such that $x^2 + ax + b = 0$ will also make $P = 0$. (See Appendix II.) Setting $x^2(x - 3)^2 + 100 = 0$, we have $x(x - 3) = 10i$ or $-10i$ where $i = \sqrt{-1}$.

By subtracting $10i$ from both sides of $x(x - 3) = 10i$, we have $x^2 - 3x - 10i = 0$, and by adding $10i$ to both sides of $x(x - 3) = -10i$, we have $x^2 - 3x + 10i = 0$.

Using the quadratic formula on each of these quadratic equations, we find $x = -1 + 2i$, or $x = 4 + 2i$, or $x = -1 - 2i$, or $x = 4 - 2i$.

So, $P = [x - (-1 + 2i)][x - (-1 - 2i)]$
$$\times [x - (4 + 2i)][x - (4 - 2i)]$$
$$= [(x + 1) - 2i][(x + 1) + 2i]$$
$$\times [(x - 4) - 2i][(x - 4) + 2i]$$
$$= (x^2 + 2x + 5)(x^2 - 8x + 20).$$

METHOD II: $x^4 - 6x^3 + 9x^2 + 100$
$$= (x^2 + ax + b)(x^2 + cx + d)$$
$$= x^4 + (a + c)x^3 + (b + ac + d)x^2 + (bc + ad)x + bd$$

$\therefore bd = 100 = 5 \cdot 20 = 10 \cdot 10 = 25 \cdot 4 = 50 \cdot 2 = 100 \cdot 1$

Try $b = 5$, $d = 20$; then $ac = -16$. Since, also, $a + c = -6$, $a = 2$ and $c = -8$ or $a = -8$ and $c = 2$. The factors are, therefore,

$$(x^2 + 2x + 5)(x^2 - 8x + 20).$$

12-20 *Express* $(a^2 + b^2)(c^2 + d^2)$ *as the sum of the squares of two binomials in four ways.*

$$
\begin{aligned}
(a^2 + b^2)(c^2 + d^2) &= a^2c^2 + a^2d^2 + b^2c^2 + b^2d^2 \\
&= a^2c^2 + 2abcd + b^2d^2 + a^2d^2 \\
&\qquad\qquad\qquad\qquad - 2abcd + b^2c^2 \\
&= (ac + bd)^2 + (ad - bc)^2
\end{aligned}
$$

NOTE: The other three forms are obtained in a similar manner. They are $(ac - bd)^2 + (ad + bc)^2$, $(ac + bd)^2 + (bc - ad)^2$, and $(bd - ac)^2 + (bc + ad)^2$.

12-21 *Observe that* 1234 *is not divisible by* 11, *but a rearrangement* (*permutation*) *of the digits such as* 1243 *is divisible by* 11. *Find the total number of permutations that are divisible by* 11.

Since divisibility by 11 requires that the difference between the sum of the odd-numbered digits and the sum of the even-numbered digits be divisible by 11 (see Appendix V), all permutations with 1 and 4 as either the odd-numbered digits or the even-numbered digits, are divisible by 11.

The number of such permutations is 8, namely, 1243, 1342, 4213, 4312, 2134, 3124, 2431, 3421.

12-22 *Find all integers* N *with initial* (*leftmost*) *digit* 6 *with the property that, when the initial digit is deleted, the resulting number is* $\frac{1}{16}$ *of the original number* N.

Let N have $k + 1$ digits. $\therefore N = 6 \cdot 10^k + y$, where y has k digits.

$\therefore 6 \cdot 10^k + y = 16y$, $y = \dfrac{6 \cdot 10^k}{15} = 4 \cdot 10^{k-1}$, with $k \geq 1$.

For $k = 1$, $y = 4$, $N = 64$.

For $k = 2$, $y = 40$, $N = 640$, and so forth.

$\therefore N = 640 \ldots 0$ with n zeros, where $n = 0, 1, 2, \ldots$.

12-23 *Find the largest positive integer that exactly divides* $N = 11^{k+2} + 12^{2k+1}$ *where* $k = 0, 1, 2, \ldots$.

By adding zero to the right side in the form of $11^k \cdot 12 - 11^k \cdot 12$, we have $N = 11^k \cdot 11^2 + 11^k \cdot 12 - 11^k \cdot 12 + (12^2)^k \cdot 12$.

$N = 11^k(11^2 + 12) + 12(144^k - 11^k)$

$N = 11^k(11^2 + 12) + 12(12^2 - 11)(144^{k-1} + \cdots + 11^{k-1})$

$N = 11^k(133) + 12(133)(144^{k-1} + \cdots + 11^{k-1})$

Since 133 appears in each term on the right, N is exactly divisible by $133 = 11^2 + 12 = 12^2 - 11$.

Challenge 1 *Find the largest positive integer exactly dividing* $N = 7^{k+2} + 8^{2k+1}$, *where* $k = 0, 1, 2, \ldots$.

Follow the solution shown above.
ANSWER: $57 = 7^2 + 8 = 8^2 - 7$

Challenge 2 *Show in general terms that* $N = A^{k+2} + (A + 1)^{2k+1}$, *where* $k = 0, 1, 2, \ldots$, *is divisible by* $(A + 1)^2 - A$.

By adding zero to the right in the form of $(A^k(A + 1) - A^k(A + 1))$, we have

$$N = A^k \cdot A^2 + A^k(A + 1) - A^k(A + 1) + ((A + 1)^2)^k(A + 1).$$
$$= A^k(A^2 + A + 1) + (A + 1)[((A + 1)^2)^k - A^k]$$
$$= A^k(A^2 + A + 1) + (A + 1)$$
$$\times [(A + 1)^2 - A][((A + 1)^2)^{k-1} + \cdots + A^{k-1}]$$
$$= A^k(A^2 + A + 1) + (A + 1)(A^2 + 2A + 1 - A)$$
$$\times [((A + 1)^2)^{k-1} + \cdots + A^{k-1}]$$

Therefore, N is divisible by $A^2 + A + 1 = (A + 1)^2 - A$ since the factor $A^2 + A + 1$ appears in each term.

12-24 *For which positive integral values of* x, *if any, is the equation* $x^6 = 9k + 1$, *where* $k = 0, 1, 2, \ldots$, *not satisfied?*

Since we are seeking multiples of 9 (increased by one), we consider those values of x that leave remainders of 0, 1, or 2 when divided by 3, since the second and higher powers of 3 are multiples of 9.

If $x = 3a + 1$, $a = 0, 1, 2, \ldots$, then $x^6 = (3a + 1)^6$. Of the seven terms in the expansion of $(3a + 1)^6$, each of the first six is divisible by 9, and the last term is 1. Therefore, $x^6 = (3a + 1)^6$ may be written as $9k + 1$.

If $x = 3a + 2$, $a = 0, 1, 2, \ldots$, we may write $x = 3b - 1$, $b = 1, 2, 3, \ldots$ where $b = a + 1$. Then $x^6 = (3b - 1)^6$. Of the seven terms in the expansion of $(3b - 1)^6$, each of the first six is divisible by 9, and the last term is 1. Therefore, $x^6 = (3b - 1)^6$ may be written as $9k + 1$.

If, however, $x = 3a$, $a = 0, 1, 2, \ldots$, then $x^6 = (3a)^6$, which, when divided by 9, leaves a remainder of 0.

Therefore, all those values of x such that $x = 3a$ where $a = 0, 1, 2, \ldots$, fail to satisfy the given equation, and all those values of x such that $x \neq 3a$ do satisfy the given equation.

12-25 *If n, A, B, and* C *are positive integers, and* $A^n - B^n - C^n$ *is divisible by* BC, *express* A *in terms of* B *and* C (*free of* n).

Using Mathematical Induction (see Appendix VII), we note that for $n = 1$, $A - B - C = k_1 BC$ so that $A = B + C + k_1 BC$ where k_1 is an integer constant. Assume that, for $n = k$, $A^k = B^k + C^k + k_2 BC$ where k_2 is an integer constant. Then, for $n = k + 1$,

$$A^{k+1} = (B^k + C^k + k_2 BC)(B + C + k_1 BC)$$
$$= B^{k+1} + C^{k+1} + (BC^k + B^k C + k_1 B^{k+1} C$$
$$+ k_1 BC^{k+1} + k_2 B^2 C + k_2 BC^2 + k_1 k_2 B^2 C^2).$$

We may write

$$BC^k + B^k C + k_1 B^{k+1} C + k_1 BC^{k+1} + k_2 B^2 C$$
$$+ k_2 BC^2 + k_1 k_2 B^2 C^2 = k_3 BC.$$

$$\therefore A^{k+1} = B^{k+1} + C^{k+1} + k_3 BC$$

Hence, BC divides $A^{k+1} - B^{k+1} - C^{k+1}$ so that the theorem is true for all natural numbers n.

$A = B + C + mBC$ where m is an integer constant.

12-26 *Prove that if* $ad = bc$, *then* $P = ax^3 + bx^2 + cx + d$, $a \neq 0$ *is divisible by* $x^2 + h^2$ *where* $h^2 = \dfrac{c}{a} = \dfrac{d}{b}$.

$$P = a\left(x^3 + \frac{b}{a}x^2 + \frac{c}{a}x + \frac{d}{a}\right), \quad d = \frac{bc}{a}, \quad c = \frac{ad}{b}$$

$$P = a\left[x\left(x^2 + \frac{c}{a}\right) + \frac{b}{a}\left(x^2 + \frac{c}{a}\right)\right], \text{ and}$$

$$P = a\left[x\left(x^2 + \frac{d}{b}\right) + \frac{b}{a}\left(x^2 + \frac{d}{b}\right)\right].$$

$$P = a\left(x^2 + \frac{c}{a}\right)\left(x + \frac{b}{a}\right), \text{ and } P = a\left(x^2 + \frac{d}{b}\right)\left(x + \frac{b}{a}\right) \text{ so}$$

that $x^2 + h^2$ divides P exactly.

COMMENT: We have proved, in addition, that, if $ad = bc$, then $-\dfrac{b}{a}$ is a root of $ax^3 + bx^2 + cx + d = 0$, $a \neq 0$. Hence, also, $-\dfrac{d}{c}$ is a root.

ILLUSTRATION: $P = 2x^3 + 4x^2 + 3x + 6$ is divisible by $x^2 + \dfrac{3}{2}$, and the real root of $P = 0$ is -2.

12-27 *Let* R *be the sum of the reciprocals of all positive factors, used once, of* N, *including* 1 *and* N, *where* $N = 2^{p-1}(2^p - 1)$ *with* $2^p - 1$ *a prime number. Find the value of* R.

The factors are $1,\ 2,\ 2^2, \ldots,\ 2^{p-1},\ 2^p - 1,\ 2(2^p - 1), \ldots,$ $2^{p-1}(2^p - 1)$:

$$R = \frac{1}{1} + \frac{1}{2} + \frac{1}{2^2} + \cdots + \frac{1}{2^{p-1}} + \frac{1}{2^p - 1}$$
$$\times \left(\frac{1}{1} + \frac{1}{2} + \frac{1}{2^2} + \cdots + \frac{1}{2^{p-1}} \right)$$
$$= \frac{2^p - 1}{2^{p-1}} + \frac{1}{2^p - 1} \cdot \frac{2^p - 1}{2^{p-1}} = \frac{2^p - 1}{2^{p-1}} \left(1 + \frac{1}{2^p - 1} \right)$$
$$= \frac{2^p - 1}{2^{p-1}} \cdot \frac{2^p}{2^p - 1} = 2,$$

where we use the formula for the sum of the terms of a geometric series. (See Appendix VII.)

12-28 *Note that* $180 = 3^2 \cdot 20 = 3^2 \cdot 2^2 \cdot 5$ *can be written as the sum of two squares of integers, namely,* $36 + 144 = 6^2 + 12^2$, *but that* $54 = 3^2 \cdot 6 = 3^2 \cdot 2 \cdot 3$ *cannot be so expressed. If* a, b *are integers, find the nature of the factor* b *such that* $a^2 \cdot b$ *is the sum of two squares of integers.*

Let $a^2 b = N^2 + M^2$.

CASE I: N and M even integers so that $N = 2K$ and $M = 2L$. Then $a^2 b = 4(K^2 + L^2) = 4r$ where $r = 0, 1, 2, \ldots$.

CASE II: N even and M odd so that $N = 2K$ and $M = 2L + 1$. Then $a^2 b = 4(K^2 + L^2 + L) + 1 = 4r + 1$.

CASE III: N and M both odd so that $N = 2K + 1$ and $M = 2L + 1$.
Then $a^2 b = 4(K^2 + K + L^2 + L) + 2 = 4r + 2$.

Therefore, when a^2b is divided by 4, the remainders are 0 or 1 or 2.

In terms of its prime factors we may write a uniquely as $a = 2^{e_1} \cdot 3^{e_2} \cdot 5^{e_3} \ldots$ (where the exponents $e_1, e_2, e_3, \ldots$ represent the number of times $2, 3, 5, \ldots$ appear as factors in a, respectively). Therefore, $a^2 = 2^{2e_1} \cdot 3^{2e_2} \cdot 5^{2e_3} \ldots$ so that, when a^2 is divided by 4, the only possible remainders are 0 or 1 (since the even powers of odd numbers leave a remainder of 1 when divided by 4, and the even powers of even numbers leave a remainder of 0 when divided by 4).

Since a^2b does not leave a remainder of 3 when divided by 4 (shown above), then b does not leave a remainder of 3 when divided by 4; that is, $b \neq 4n + 3$, where $n = 0, 1, 2, \ldots$, in order for a^2b to represent the sum of two squares of integers.

12-29 *Show that* $b - 1$ *divides* $b^{b-2} + b^{b-3} + \cdots + b + 1$, *and thus show that* $b^2 - 2b + 1$ *divides* $b^{b-1} - 1$.

PROOF I: Let $N = b^{b-2} + b^{b-3} + \cdots + b + 1$, and interpret N as a number in base b with $b - 1$ digits, each a 1. Thus, N is divisible by $b - 1$ since a number in base b is divisible by $b - 1$ if the sum of its digits is divisible by $b - 1$. (See Problem 12-3.)

Since $b^{b-1} - 1 = (b - 1)(b^{b-2} + b^{b-3} + \cdots + b + 1) = (b - 1) N$, and N is divisible by $b - 1$, then $b^{b-1} - 1$ is divisible by $(b - 1)^2 = b^2 - 2b + 1$.

PROOF II: Let

$$S = b^{b-2} + b^{b-3} + \cdots + b + 1$$
$$= [(b^{b-2} - 1) + 1] + [(b^{b-3} - 1) + 1] + \cdots$$
$$+ [(b - 1) + 1] + 1$$
$$= (b^{b-2} - 1) + (b^{b-3} - 1) + \cdots + (b - 1) +$$
$$\underbrace{1 + 1 + \cdots + 1}_{b - 1}$$
$$= (b^{b-2} - 1) + (b^{b-3} - 1) + \cdots + (b - 1) + (b - 1)$$

Since each term on the right is divisible by $b - 1$, S is divisible by $b - 1$. The proof concludes following the reasoning given in the second paragraph of Proof I.

13 Maxima and Minima: Ups and Downs

13-1 *The perimeter of a sector of a circle is 12 (units). Find the radius so that the area of the sector is a maximum.*

Let K represent the area of the sector. Then $K = \frac{1}{2} rs$ where r is the radius and s is the arc-length. Since $r + r + s = 12$, $s = 12 - 2r$, so $K = \frac{1}{2} r(12 - 2r) = 6r - r^2$. Therefore, $K = 9 - (9 - 6r + r^2) = 9 - (3 - r)^2$ (completing the square and factoring). K is maximum when $r = 3$ (units), since K equals 9 when $(3 - r)^2 = 0$, or less when $(3 - r)^2 \neq 0$.

13-2 *The seating capacity of an auditorium is 600. For a certain performance, with the auditorium not filled to capacity, the receipts were \$330.00. Admission prices were 75¢ for adults and 25¢ for children. If a represents the number of adults at the performance, find the minimum value of a satisfying the given conditions.*

Let c represent the number of children. Then $a + c < 600$,

$$\text{and } \frac{1}{4}a + \frac{1}{4}c < 150. \tag{I}$$

$$\text{We know that } \frac{3}{4}a + \frac{1}{4}c = 330. \tag{II}$$

Subtracting I from II, we find that $\frac{1}{2}a > 180$ and $a > 360$. Therefore, the minimum value for a is 361.

13-3 *When the admission price to a ball game is 50 cents, 10,000 persons attend. For every increase of 5 cents in the admission price, 100 fewer (than the 10,000) attend. Find the admission price that yields the largest income.*

Represent the admission price yielding the largest income by $50 + 5n$. Then the income, in dollars, becomes

$$I = \left(\frac{50 + 5n}{100}\right)(10{,}000 - 200n).$$

$I = 5000 + 400n - 10n^2 = 10[900 - (20 - n)^2]$ so that I (maximum) occurs when $n = 20$. The required admission price is $50 + 5 \cdot 20 = 150$ (cents), or \$1.50.

Challenge 3 *Find the admission price yielding the largest income if, in addition to the conditions stated in the original problem, there is an additional expense of one dollar for every 100 persons in attendance.*

$I = \left(\dfrac{50 + 5n}{100}\right)(10{,}000 - 200n) - (10{,}000 - 200n)$. The answer is $50 + 150 = 200$ (cents), or \$2.00.

13-4 *A rectangle is inscribed in an isosceles triangle with base 2b (inches) and height h (inches), with one side of the rectangle lying in the base of the triangle. Let T (square inches) be the area of the triangle, and R_m the area of the largest rectangle so inscribed. Find the ratio R_m:T.*

Designate the base of the rectangle as $2x$ and the altitude as y. From similar triangles we have the proportion $\dfrac{y}{h} = \dfrac{b - x}{b}$ so that $x = \dfrac{b}{h}(h - y)$. Since $R = 2xy$, we have

$$R = 2y \cdot \frac{b}{h}(h - y) = \frac{2b}{h}(hy - y^2). \qquad (I)$$

Adding to the right side of equation I zero in the form $-\dfrac{2b}{h}\left(\dfrac{h^2}{4}\right) + \dfrac{bh}{2}$, we get $R = \dfrac{2b}{h}\left(-\dfrac{h^2}{4} + hy - y^2\right) + \dfrac{bh}{2}$. Therefore, $R = \dfrac{bh}{2} - \dfrac{2b}{h}\left(y - \dfrac{h}{2}\right)^2$.

R_m, the maximum value of R, is $\dfrac{bh}{2}$, obtained when $y = \dfrac{h}{2}$. Since $T = \dfrac{1}{2}(2b)(h) = bh$, the ratio R_m:$T = 1$:2.

13-5 *It can be proved that the function $f(y) = ay - y^b$ where $b > 1$, $a > 0$, and $y \geq 0$, takes its largest value when $y = \left(\dfrac{a}{b}\right)^{\frac{1}{b-1}}$. Use this theorem to find the maximum value of the function $F = \sin x \sin 2x$.*

$F = \sin x \sin 2x = \sin x(2 \sin x \cos x)$. For $\sin^2 x$ we substitute $1 - \cos^2 x$ and obtain $F = 2(\cos x - \cos^3 x)$. By letting $y = \cos x$, we convert F into $F = 2(y - y^3)$.

To maximize the function $y - y^3$, we note, by comparing it to $f(y)$, that $a = 1$ and $b = 3$. Hence, $y - y^3$ will take its largest value when $y = \left(\frac{1}{3}\right)^{\frac{1}{3-1}} = \frac{1}{\sqrt{3}}$. Therefore, $F(\max) = 2\left(\frac{1}{\sqrt{3}} - \left(\frac{1}{\sqrt{3}}\right)^3\right) = \frac{4}{3\sqrt{3}}$.

13-6 *In the woods 12 miles north of a point* B *on an east-west road, a house is located at point* A. *A power line is to be built to* A *from a station at* E *on the road, 5 miles east of* B. *The line is to be built either directly from* E *to* A *or along the road to a point* P (*between* E *and* B), *and then through the woods from* P *to* A, *whichever is cheaper. If it costs twice as much per mile building through the woods as it does building along the highway, find the location of point* P *with respect to point* B *for the cheapest construction.*

Represent the cost function by C, and the distance from B to P by x. Then $C = 5 - x + 2\sqrt{144 + x^2}$. With the aid of a table of square roots, graph the given function for $0 \leq x \leq 5$. Minimum C occurs when $x = 5$, so that P is at E, that is, 5 miles east of B.

13-7 *From a rectangular cardboard 12 by 14, an isosceles trapezoid and a square, of side length* s, *are removed so that their combined area is a maximum. Find the value of* s.

Let A represent the combined area. (See Fig. S13-7.) Then, since $A = \frac{1}{2} h(b_1 + b_2) + s^2$, where h is the altitude of the trapezoid and b_1 and b_2 are its bases, and s is the side of the square, $A = \frac{1}{2}(12 - s)(14 + s) + s^2 = \frac{1}{2}s^2 - s + 84$.

In order to determine the maximum value of A more readily, we rewrite it as $A = \frac{1}{2}(s^2 - 2s + 1) + 84 - \frac{1}{2} = \frac{1}{2}(s - 1)^2 + 83\frac{1}{2}$. Obviously, A is a maximum when s is a maximum, that is,

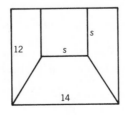

S13-7

when $s = 12$. For this value of s, the area of the square is 144 and the area of the trapezoid is zero.

COMMENT: Note that the minimum value of the area is $83\frac{1}{2}$, occurring when $s = 1$, but that the combined area is 84 when $s = 0$.

13-8 *Two equilateral triangles are to be constructed from a line segment of length* L. *Determine their perimeters* P_1 *and* P_2 *so that* **(a)** *the combined area is a maximum* **(b)** *the combined area is a minimum.*

If we represent the perimeter of one triangle by P_1, the perimeter of the second triangle is represented by $L - P_1$. Therefore, the combined area $A = \frac{1}{4}\left(\frac{P_1}{3}\right)^2 \sqrt{3} + \frac{1}{4}\left(\frac{L - P_1}{3}\right)^2 \sqrt{3}$ (using the formula $\frac{1}{4}s^2 \sqrt{3}$ for the area of an equilateral triangle). Hence,

$$A = \frac{\sqrt{3}}{36}(P_1{}^2 + L^2 - 2LP_1 + P_1{}^2) = \frac{\sqrt{3}}{36}[2(P_1{}^2 - LP_1) + L^2].$$

Adding to the right side zero in the form of $2 \cdot \frac{L^2}{4} - \frac{L^2}{2}$, we obtain

$$A = \frac{\sqrt{3}}{36}\left[2\left(P_1 - \frac{L}{2}\right)^2 + L^2 - \frac{L^2}{2}\right] = \frac{\sqrt{3}}{36}\left[2\left(P_1 - \frac{L}{2}\right)^2 + \frac{L^2}{2}\right].$$

Since the least value for $\left(P_1 - \frac{L}{2}\right)^2$ is zero, a value obtained when $P_1 = \frac{L}{2}$, A is minimum when $P_1 = \frac{L}{2}$ so that the minimum confined area occurs when $P_1 = \frac{L}{2}$ and $P_2 = L - \frac{L}{2} = \frac{L}{2}$; that is, $P_1 = P_2 = \frac{L}{2}$.

Since $P_1 \geq 0$ we find, by inspection, that A is maximum when $P_1 = 0$ so that $P_2 = L$, or $P_2 = 0$ so that $P_1 = L$; that is, the maximum combined area occurs when all of L is used for just one triangle.

COMMENT: Note that the maximum A equals twice minimum A.

13-9 *Find the least value of* $x^4 + y^4$ *subject to the restriction* $x^2 + y^2 = c^2$.

$$\text{Let } F = x^4 + y^4 = (x^2 + y^2)^2 - 2x^2y^2. \tag{I}$$

Since $2x^2y^2 \geq 0$, F is obviously least when $2x^2y^2$ is greatest, and $2x^2y^2$ is greatest when x^2y^2 is greatest.

Since $x^2 + y^2 = c^2$, x^2y^2 is greatest when $x^2 = y^2 = \frac{c^2}{2}$.

(If the sum of two numbers is a constant, their product is greatest when each is half the constant.) Therefore, $x^2y^2(\text{max}) = \dfrac{c^2}{2} \cdot \dfrac{c^2}{2} = \dfrac{c^4}{4}$, and so $2x^2y^2(\text{max}) = 2 \cdot \dfrac{c^4}{4} = \dfrac{c^4}{2}$.

Therefore, from (I), $F(\text{min}) = (c^2)^2 - \dfrac{c^4}{2} = \dfrac{c^4}{2}$.

Challenge *Find the least value of* $x^3 + y^3$ *subject to the restriction* $x + y = c$.

Follow the pattern of the original solution. The answer is $\dfrac{c^3}{4}$.

13-10 *Find the value of* x *such that* $S = (x - k_1)^2 + (x - k_2)^2 + \cdots + (x - k_n)^2$ *is a minimum where each* k_i, $i = 1, 2, \ldots, n$, *is a constant.*

METHOD I: Using the symbol $\sum$ for summation we write

$$S = \sum_{i=1}^{n} (x - k_i)^2 = \sum_{i=1}^{n} (x^2 - 2xk_i + k_i^2).$$

When written out, the terms of S are $(x^2 - 2xk_1 + k_1^2) + (x^2 - 2xk_2 + k_2^2) + (x^2 - 2xk_3 + k_3^2) + \cdots + (x^2 - 2xk_n + k_n^2)$.

Therefore, $S = (x^2 + x^2 + x^2 + \cdots) - 2x(k_1 + k_2 + \cdots + k_n) + (k_1^2 + k_2^2 + \cdots + k_n^2)$.

Since, in the complete expansion of S, the term x^2 appears n times, $S = nx^2 - 2x\sum k_i + \sum k_i^2$. To the right side we add zero in the form of $(n) \dfrac{(\sum k_i)^2}{n^2} - \dfrac{(\sum k_i)^2}{n}$ so that

$$S = n \left(x^2 - 2x \frac{\sum k_i}{n} + \frac{(\sum k_i)^2}{n^2} \right) + \sum k_i^2 - \frac{(\sum k_i)^2}{n}.$$

Hence, $S = n \left(x - \dfrac{\sum k_i}{n} \right)^2 + \sum k_i^2 - \dfrac{(\sum k_i)^2}{n}$.

The minimum value of S occurs when $\left(x - \dfrac{\sum k_i}{n} \right)^2$ equals zero (see Problem 13-8); that is, when $x = \dfrac{\sum k_i}{n}$, which, interpreted, means that x is the arithmetic mean of the k_i.

METHOD II: Compare the expression $S = nx^2 - 2x\sum k_i + \sum k_i^2$ with that of $y = ax^2 + bx + c$, the equation of a parabola. As minimum y occurs when $x = -\dfrac{b}{2a}$, so minimum S occurs when $x = -\dfrac{-2\sum k_i}{2n} = \dfrac{\sum k_i}{n}$.

METHOD III: If you are familiar with the calculus, solve the problem with the use of the first derivative.

ILLUSTRATION: $S = (x - 2)^2 + (x + 3)^2$;

$$x = \frac{\sum k_i}{2} = \frac{2 + (-3)}{2} = -\frac{1}{2}$$

$$S(\min) = \frac{25}{4} + \frac{25}{4} = 12\frac{1}{2}$$

$$S(\min) = \sum k_i^2 - \frac{(\sum k_i)^2}{n} = 13 - \frac{(-1)^2}{2} = 13 - \frac{1}{2} = 12\frac{1}{2}$$

13-11 *If* $|x| \leq c$ *and* $|x - x_1| \leq 1$, *find the greatest possible value of* $|x_1{}^2 - x^2|$.

$|x_1{}^2 - x^2| = |x_1 + x| \, |x_1 - x|$.

Since $|x_1 - x| \leq 1$,

$|x_1{}^2 - x^2| \leq |x_1 + x| = |2x + x_1 - x|$.

Therefore, $|x_1{}^2 - x^2| \leq |2x| + |x_1 - x| \leq 2c + 1$.

A geometric interpretation is helpful (Fig. S13-11). Since $|x_1 - x| \leq 1$, $x_1 \leq x + 1$. Therefore, $x_1 + x \leq 2x + 1 \leq 2c + 1$. The maximum value of $|x_1{}^2 - x^2| = |x_1 + x| \, |x_1 - x|$ is represented by the area of rectangle $ABCD$ with base $DC = 2c + 1$ and altitude $CB = 1$.

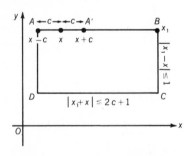

S13-11

13-12 *Show that the maximum value of* $F = \dfrac{ab}{4(a + b)^2}$, *where* a, b *are positive numbers, is* $\dfrac{1}{16}$.

When $a = b$, $F = \dfrac{a^2}{16a^2} = \dfrac{1}{16}$;

when $a \neq b$, $\dfrac{a + b}{2} > \sqrt{ab}$. (See Appendix IV.) Therefore,

$\dfrac{(a + b)^2}{4} > ab$ and, hence, $\dfrac{ab}{(a + b)^2} < \dfrac{1}{4}$.

Since $F = \dfrac{ab}{4(a + b)^2}$, $F < \dfrac{1}{4} \cdot \dfrac{1}{4} = \dfrac{1}{16}$.

Therefore, $F(\text{max}) = \dfrac{1}{16}$.

13-13 *Find the area of the largest trapezoid that can be inscribed in a semicircle of radius* r.

METHOD I: The upper base and the (equal) legs of the trapezoid may be considered three sides of a hexagon inscribed in the circle. Since, of all inscribed hexagons, the regular hexagon has maximum area, the trapezoid of maximum area is the one where the upper base (and each leg) equals a side of the regular hexagon, that is, where $2x = r$. The lower base, of course, is the diameter. This value of x gives an area of $\dfrac{3\sqrt{3}}{4} r^2$, a maximum (i.e., the area of three equilateral triangles of side length r).

METHOD II: $K = \dfrac{1}{2}(2r + 2x)(r^2 - x^2)^{\frac{1}{2}}$. Take $r = 1$; then $K = (1 + x)(1 - x^2)^{\frac{1}{2}}$. Using a table of square roots, if necessary, plot the graph of K for $0 \leq x \leq 1$ (Fig. S13-13). It is found that maximum K occurs when $x = \dfrac{1}{2}$, and so $K(\text{max}) = \dfrac{3\sqrt{3}}{4}$.

Generally, $K(\text{max}) = \dfrac{3\sqrt{3}}{4} r^2$ where r is the radius of the circle.

METHOD III: Use the calculus, as shown in the solution of Problem 13-10, Method III.

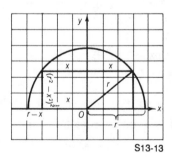

S13-13

14 Quadratic Equations: Fair and Square

14-1 *Find the real values of* x *such that* $3^{2x^2-7x+3} = 4^{x^2-x-6}$.

With the exponents in factored form we get: $3^{(2x-1)(x-3)} = 4^{(x+2)(x-3)}$. If $x = 3$, both sides are equal to 1, and we have a root. If $x \neq 3$, we can take the $(x-3)$rd root of both sides to get: $3^{(2x-1)} = 4^{(x+2)}$.

Therefore $(2x-1)\log 3 = (x+2)\log 4$. Thus, $x = \dfrac{1+2\left\{\dfrac{\log 4}{\log 3}\right\}}{2-\left\{\dfrac{\log 4}{\log 3}\right\}}$

14-2 *Let* $D = h^2 + 3k^2 - 2hk$ *where* h, k, *are real numbers. For what values of* h *and* k *is* D > 0?

$D = h^2 + 3k^2 - 2hk = h^2 - 2hk + k^2 + 2k^2 = (h - k)^2 + 2k^2$. Since $(h - k)^2 \geq 0$ and $2k^2 \geq 0$, $D > 0$ for all real values of h, k except $h = k = 0$.

14-3 *If the roots of* $x^2 + bx + c = 0$ *are the squares of the roots of* $x^2 + x + 1 = 0$, *find the values of* b *and* c.

Since the roots of $x^2 + x + 1 = 0$ are squares of each other (show this!), $b = 1$ and $c = 1$.

14-4 *If the roots of* $ax^2 + bx + c = 0$, $a \neq 0$, *are in the ratio* m:n, *find an expression relating* m *and* n *to* a, b, *and* c.

Let $\dfrac{r_1}{r_2} = \dfrac{m}{n}$; then $r_1 = km$, $r_2 = kn$, where r_1 and r_2 are the roots of the given equation. Since $r_1 + r_2 = -\dfrac{b}{a}$ and $r_1 r_2 = \dfrac{c}{a}$, then $\left(-\dfrac{b}{a}\right)^2 = k^2(m + n)^2$ and $\dfrac{c}{a} = k^2 mn$. Therefore, $\dfrac{\left(-\dfrac{b}{a}\right)^2}{\dfrac{c}{a}} = \dfrac{k^2(m + n)^2}{k^2 mn}$, and so $mnb^2 = (m + n)^2 ac$.

A second form of the relation sought is $\dfrac{b^2}{ac} = \dfrac{m}{n} + 2 + \dfrac{n}{m} = \left(\sqrt{\dfrac{m}{n}} + \sqrt{\dfrac{n}{m}}\right)^2$.

NOTE: For equal roots, the discriminant $b^2 - 4ac = 0$ or $b^2 = 4ac$. With $m = n = 1$, the first form becomes $1 \cdot 1 \cdot b^2 = (1 + 1)^2 ac$; that is, $b^2 = 4ac$.

ILLUSTRATION: For $m = 3$, $n = 2$, $b = 5$, $6 \cdot 5^2 = 5^2 ac$, so that $a = 6$, $c = 1$, or $a = 1$, $c = 6$, or $a = 3$, $c = 2$ or $a = 2$, $c = 3$. For example, the roots of $2x^2 + 5x + 3 = 0$ are $-\dfrac{3}{2}$ and -1 with a ratio $3:2$.

14-5 *Find all values of* x *satisfying the pair of equations* $x^2 - px + 20 = 0$, $x^2 - 20x + p = 0$.

CASE I: When $p = 20$, the equations are identical and satisfied by two values of x, $10 + 4\sqrt{5}$ and $10 - 4\sqrt{5}$.

CASE II: When $p \ne 20$, then $x^2 - px + 20 = x^2 - 20x + p$, $(20 - p)x = p - 20$. Since $p \ne 20$, we obtain by division $x = -1$.

To satisfy the given equations with the value $x = -1$, the value of p must be -21.

Generally, the pair of equations $x^2 - px - p - 1 = 0$ and $x^2 + (p + 1)x + p = 0$ is satisfied by $x = -1$.

14-6 *A student, required to solve the equation* $x^2 + bx + c = 0$, *inadvertently solves the equation* $x^2 + cx + b = 0$; b, c *integers. One of the roots obtained is the same as a root of the original equation, but the second root is* m *less than the second root of the original equation. Find* b *and* c *in terms of* m.

Let the roots of the original equation be r, s. Then $r + s = -b$ and $r + s - m = -c$.

$$r + s = m - c; \quad -b = m - c; \quad c - b = m \qquad \text{(I)}$$
$$rs = c \text{ and } rs - rm = b;$$
$$rs = b + rm; \quad c = b + rm; \quad c - b = rm \qquad \text{(II)}$$

From (I) and (II), $m = rm$ and $r = 1$.

$rs = s = c$ and $r + s = 1 + s = -b$, or $s = -b - 1 = c$, so $c + b = -1$.

Since $c - b = m$ (I), $c = \dfrac{m - 1}{2}$, $b = \dfrac{-m - 1}{2}$.

14-7 *If* r_1 *and* r_2 *are the roots of* $x^2 + bx + c = 0$, *and* $S_2 = r_1^2 + r_2^2$, $S_1 = r_1 + r_2$, *and* $S_0 = r_1^0 + r_2^0$, *prove that* $S_2 + bS_1 + cS_0 = 0$.

METHOD I: Since $S_2 = r_1^2 + r_2^2 = (r_1 + r_2)^2 - 2r_1r_2$, then $S_2 = (-b)^2 - 2c = b^2 - 2c$, since $r_1 + r_2 = -b$ and $r_1r_2 = c$.

$$\therefore S_2 + bS_1 + cS_0 = b^2 - 2c + b(-b) + c \cdot 2 = 0$$

METHOD II: $S_2 + bS_1 + cS_0$
$$= r_1^2 + r_2^2 + b(r_1 + r_2) + c(r_1^0 + r_2^0)$$
$$= r_1^2 + br_1 + c + r_2^2 + br_2 + c$$

Since r_1 is a root of the given equation, $r_1^2 + br_1 + c = 0$.
Similarly, $r_2^2 + br_2 + c = 0$.
Consequently, $S_2 + bS_1 + cS_0 = 0$.

14-8 *A man sells a refrigerator for* $171, *gaining on the sale as many percent (based on the cost) as the refrigerator cost,* C, *in dollars. Find* C.

Since Selling Price = Cost plus Profit, $171 = C + \left(\frac{C}{100}\right)C$,
$C^2 + 100C = 17{,}100$. By completing the square, we get $C^2 + 100C + 2500 = 17{,}100 + 2500$, $(C + 50)^2 = 140^2$, $C + 50 = 140$, $C = 90$. $C^2 + 100C - 17{,}100 = 0$ implies $(C - 90) \times (C + 190) = 0$, and, hence, $C = 90$.

14-9 *Express* q *and* s *each in terms of* p *and* r *so that the equation* $x^4 + px^3 + qx^2 + rx + s = 0$ *has two double roots* u *and* v *where* u *may or may not equal* v. (*Each of the factors* x − u *and* x − v *appears twice in the factorization of* $x^4 + px^3 + qx^2 + rx + s$.)

Since each of the factors $x - u$ and $x - v$ appears twice, we may write $x^4 + px^3 + qx^2 + rx + s = (x - u)(x - v)(x - u) \times (x - v) = (x^2 - x(u + v) + uv)^2$. Let $u + v = -b$ and let $uv = c$; then $x^2 - x(u + v) + uv = x^2 + bx + c$.
Therefore, $x^4 + px^3 + qx^2 + rx + s = (x^2 + bx + c)^2 = x^4 + 2bx^3 + (b^2 + 2c)x^2 + 2bcx + c^2$. $\qquad$ (I)

Since (I) is an identity in x, we may equate the coefficients of like powers of x on both sides of the identity, and, therefore, $p = 2b$, $r = 2bc$ so that $\dfrac{r}{p} = c$.

Since $s = c^2$ (from I), $s = \dfrac{r^2}{p^2}$, and since $q = b^2 + 2c$ (from I), $q = \dfrac{p^2}{4} + \dfrac{2r}{p}$.

14-10 *Let* $f(n) = n(n + 1)$ *where* n *is a natural number. Find values of* n *such that* $f(n + 4) = 4f(n) + 4$.

Since $f(n + 4) = 4f(n) + 4$, $(n + 4)(n + 5) = 4n(n + 1) + 4$. Therefore, $3n^2 - 5n - 16 = 0$. This equation has no solution in integers. Therefore, no value of n satisfies the required condition.

14-11 *If one root of* $Ax^3 + Bx^2 + Cx + D = 0$, $A \neq 0$, *is the arithmetic mean of the other two roots, express the simplest relation between* A, B, C, *and* D.

Let r_1, r_2, r_3 be the roots of the given equation with r_2 the arithmetic mean of r_1 and r_3. Therefore, $2r_2 = r_1 + r_3$. Since $r_1 + r_2 + r_3 = -\dfrac{B}{A}$, a generalization for the sum of the roots, $3r_2 = -\dfrac{B}{A}$ so that $r_2 = -\dfrac{B}{3A}$.

Since r_2 is a root of the given equation, $Ar_2{}^3 + Br_2{}^2 + Cr_2 + D = 0$. Therefore, $A\left(-\dfrac{B}{3A}\right)^3 + B\left(-\dfrac{B}{3A}\right)^2 + C\left(-\dfrac{B}{3A}\right) + D = 0$. Simplifying, we obtain $2B^3 - 9ABC + 27A^2D = 0$.

Challenge *Find the simplest relation between the coefficients if one root is the positive geometric mean of the other two.*

Let the root r_2 be the positive geometric mean of the other two roots r_1 and r_3, that is, let $r_2 = \sqrt{r_1 r_3}$. Therefore, $r_2{}^2 = r_1 r_3$ and $r_2{}^3 = r_1 r_2 r_3$.

Since $r_1 r_2 r_3 = -\dfrac{D}{A}$, a generalization for the product of the roots, $r_2{}^3 = -\dfrac{D}{A}$ so that $r_2 = \sqrt[3]{-\dfrac{D}{A}}$. Substituting this value for r_2 back into the original equation, we have

$$A\left(\sqrt[3]{-\dfrac{D}{A}}\right)^3 + B\left(\sqrt[3]{-\dfrac{D}{A}}\right)^2 + C\left(\sqrt[3]{-\dfrac{D}{A}}\right) + D = 0.$$

After simplification, this last equation yields $B^3D - AC^3 = 0$.

14-12 *If the coefficients* a, b, c *of the equation* $ax^2 + bx + c = 0$ *are odd integers, find a relation between* a, b, c *for which the roots are rational.*

We prove that if a, b, c are odd integers, there are no rational roots.

METHOD I: For rational roots, the discriminant must be the square of an integer. Let $b^2 - 4ac = t^2$, and $b^2 - t^2 = 4ac = (b - t)(b + t)$ where both b and t are odd. (Why is t odd?) Let $b = 2b_1 + 1$ and let $t = 2t_1 + 1$.

Therefore, $(2b_1 + 1 - 2t_1 - 1)(2b_1 + 1 + 2t_1 + 1) = 4ac$,

$$2(b_1 - t_1)(2)(b_1 + t_1 + 1) = 4ac,$$

and $(b_1 - t_1)(b_1 + t_1 + 1) = ac$.

The product ac is odd. If b_1 and t_1 are each even, then $b_1 - t_1$ is even. If b_1 and t_1 are each odd, then $b_1 - t_1$ is even. If one of b_1, t_1 is odd and the other is even, then $b_1 + t_1 + 1$ is even. Hence, for all possibilities, $(b_1 - t_1)(b_1 + t_1 + 1)$ is even; we have a contradiction.

METHOD II: Let $b = 2b_1 + 1$, $a = 2a_1 + 1$, $c = 2c_1 + 1$. Then $D = b^2 - 4ac = (2b_1 + 1)^2 - 4(2a_1 + 1)(2c_1 + 1) = 8\left[\dfrac{b_1(b_1 + 1)}{2} - 2a_1c_1 - a_1 - c_1 - 1\right] + 5$. Since $\dfrac{b_1(b_1 + 1)}{2}$ is integral, then $D = 8k + 5$. If $D = N^2$, with N odd, then, $N^2 = 8k + 5$. However, the square of an odd number leaves a remainder of 1 when divided by 8; that is, $(4k \pm 1)^2 \equiv 1 \pmod 8$.

Therefore, rational roots are impossible.

14-13 *If* $f(x) = a_0x^2 + a_1x + a_2 = 0$, $a_0 \neq 0$, *and* a_0, a_2, *and* $s = a_0 + a_1 + a_2$ *are odd numbers, prove that* $f(x) = 0$ *has no rational root.*

It is not much more difficult to prove the more general theorem for an equation of degree n, of which this is the special case with $n = 2$. If $f(x) = a_0x^n + a_1x^{n-1} + \cdots + a_n = 0$, $a_0 \neq 0$, and a_0, a_n, and $s = a_0 + a_1 + a_2 + \cdots + a_n$ are odd numbers, prove that $f(x) = 0$ has no rational root.

PROOF: (1) Assume that $f(x) = 0$ does have a rational root $\dfrac{p}{q}$ expressed in lowest terms. Then, since a_n is odd, p must be odd, and, since a_0 is odd, q must be odd.

(2) We have $a_0 \dfrac{p^n}{q^n} + a_1 \dfrac{p^{n-1}}{q^{n-1}} + \cdots + a_{n-1} \dfrac{p}{q} + a_n = 0$. Clearing the equation of fractions, we get $a_0 p^n + a_1 p^{n-1} q + \cdots + a_{n-1} p q^{n-1} + a_n q^n = 0$.

(3) Also, $a_0 + a_1 + \cdots + a_{n-1} + a_n = $ odd number (hypothesis).

(4) $\therefore S = a_0(1 + p^n) + a_1(1 + p^{n-1}q) + a_2(1 + p^{n-2}q) + \cdots + a_{n-1}(1 + pq^{n-1}) + a_n(1 + q^n)$ should be equal to an odd number.

(5) But $1 + p^n$ is even, $1 + p^{n-1}q$ is even, ..., $1 + q^n$ is even, so that S is even.

(6) This contradiction shows that our assumption of a rational root was false.

The only possibility left is that $f(x) = 0$ has no rational root.

---

15 Systems of Equations: Strictly Simultaneous

15-1 *Estimate the values of the four variables in the given linear system. Then substitute repeatedly until a definitive solution is reached.*

$$x_1 = \frac{1}{4}(0 + x_2 + x_3 + 0)$$

$$x_2 = \frac{1}{4}(0 + 0 + x_4 + x_1)$$

$$x_3 = \frac{1}{4}(x_1 + x_4 + 1 + 0)$$

$$x_4 = \frac{1}{4}(x_2 + 0 + 1 + x_3)$$

We begin by attempting a "reasonable guess" at the values of x_1, x_2, x_3, and x_4, assuming that such values exist.

To keep things simple let us guess $x_1 = x_2 = x_3 = x_4 = \dfrac{1}{4}$. Substituting into the given equations, we find

$$x_1 = \frac{1}{4}\left(0 + \frac{1}{4} + \frac{1}{4} + 0\right) = \frac{1}{8}.$$

Similarly, $x_2 = \frac{1}{4}\left(0 + 0 + \frac{1}{4} + \frac{1}{4}\right) = \frac{1}{8}$,

$$x_3 = \frac{1}{4}\left(\frac{1}{4} + \frac{1}{4} + 1 + 0\right) = \frac{3}{8}, \text{ and}$$

$$x_4 = \frac{1}{4}\left(\frac{1}{4} + 0 + 1 + \frac{1}{4}\right) = \frac{3}{8}.$$

We now repeat the process seeking to "refine" the second set of values found for the unknowns. It happens that no further refinement occurs, as we illustrate with x_1:

$x_1 = \frac{1}{4}\left(0 + \frac{1}{8} + \frac{3}{8} + 0\right) = \frac{1}{8}$, the value we already have. The values $x_1 = x_2 = \frac{1}{8}$ and $x_3 = x_4 = \frac{3}{8}$ are, then, the exact values since they satisfy the given equations.

We happened to start with a fortunate guess. What if our original guess was $x_1 = x_2 = x_3 = x_4 = \frac{1}{2}$? We do obtain the values $x_1 = x_2 = \frac{1}{8}, x_3 = x_4 = \frac{3}{8}$, but not until six "feedbacks." The first yields $x_1 = x_2 = \frac{1}{4}$, $x_3 = x_4 = \frac{1}{2}$; the second, $x_1 = x_2 = \frac{3}{16}$, $x_3 = x_4 = \frac{7}{16}$; the third, $x_1 = x_2 = \frac{5}{32}$, $x_3 = x_4 = \frac{13}{32}$; the fourth, $x_1 = x_2 = \frac{9}{64}$, $x_3 = x_4 = \frac{25}{64}$; the fifth, $x_1 = x_2 = \frac{17}{128}$, $x_3 = x_4 = \frac{49}{128}$; the sixth, $x_1 = x_2 = \frac{1}{8}$, $x_3 = x_4 = \frac{3}{8}$.

COMMENT: The chief value in this method lies in finding values of the unknowns that are approximate, involving, perhaps, several decimal places.

15-2 *For the system* x + y + 2z = a, (I)

$$-2x - z = b \quad\quad\quad \text{(II)}$$

$$x + 3y + 5z = c \quad\quad\quad \text{(III)}$$

find a relation between a, b, *and* c *so that a solution exists other than* x = 0, y = 0, z = 0.

Note that, if $a = b = c = 0$, the system is satisfied for $x = y = z = 0$, but this solution, referred to as the trivial solution, is here ruled out. We seek other solutions if they exist.

We find that the value of the system determinant

$$D = \begin{vmatrix} 1 & 1 & 2 \\ -2 & 0 & -1 \\ 1 & 3 & 5 \end{vmatrix} = 0.$$

For the system to be determined, that is, for solutions to exist (Cramer's Rule, Appendix VII), $D_x = D_y = D_z = 0$ where

$$D_x = \begin{vmatrix} a & 1 & 2 \\ b & 0 & -1 \\ c & 3 & 5 \end{vmatrix}, \qquad D_y = \begin{vmatrix} 1 & a & 2 \\ -2 & b & -1 \\ 1 & c & 5 \end{vmatrix},$$

$$D_z = \begin{vmatrix} 1 & 1 & a \\ -2 & 0 & b \\ 1 & 3 & c \end{vmatrix}$$

Expanding D_x, we have $D_x = -c + 6b - 5b + 3a$. Therefore, $3a + b - c = 0$, and this is the relation between a, b, and c that assures non-trivial solutions. The same result is obtained when D_y is expanded and when D_z is expanded.

COMMENT: To find the solutions, of which there are infinitely many, providing $3a + b - c = 0$, you may choose the value of x arbitrarily. Eliminate z from equations I and II and obtain $y = 3x - 5a + 2c$, after replacing b by its equivalent $c - 3a$. The value of z can then be found from equation I or equation III to be $z = -2x + 3a - c$.

To illustrate, let $x = 1$, then $y = 3 - 5a + 2c$ and $z = -2 + 3a - c$. This set of values satisfies the given system, as you can verify for yourself, keeping in mind that $b = c - 3a$.

15-3 *Find the smallest value of* p^2 *for which the pair of equations,*

$$(4 - p^2)x + 2y = 0,$$
$$2x + (7 - p^2)y = 0$$

has a solution other than $x = y = 0$, *and find the ratio* $x:y$ *for this value of* p^2.

Multiply $(4 - p^2)x + 2y = 0$ by $7 - p^2$ to obtain

$$(4 - p^2)(7 - p^2)x + 2y(7 - p^2) = 0, \tag{I}$$

and multiply $2x + (7 - p^2)y = 0$ by 2 to obtain

$$4x + 2y(7 - p^2) = 0. \tag{II}$$

By subtracting (II) from (I) we obtain

$$[(4 - p^2)(7 - p^2) - 4]x = 0.$$

Since $x \neq 0$, $(4 - p^2)(7 - p^2) - 4 = 0$; that is, $p^4 - 11p^2 + 24 = 0$.

Since $p^4 - 11p^2 + 24 = (p^2 - 3)(p^2 - 8)$, $p^2 - 3 = 0$ and, therefore, $p^2 = 3$. The value $p^2 = 8$ is rejected since $8 > 3$.

Substituting $p^2 = 3$ into the second of the original equations, we have $2x + 4y = 0$, so that $x:y = -2;1$.

COMMENT: For the value $p^2 = 8$, the ratio $x:y = 1:2$.

15-4 *If* $P_1 = 2x^4 + 3x^3 - 4x^2 + 5x + 3$,
   $P_2 = x^3 + 2x^2 - 3x + 1$,
   $P_3 = x^4 + 2x^3 - x^2 + x + 2$,
and $aP_1 + bP_2 + cP_3 = 0$, *find the value of* a + b + c *where* abc $\neq 0$.

$aP_1 + bP_2 + cP_3 = (2a + c)x^4 + (3a + b + 2c)x^3 +$
$(-4a + 2b - c)x^2 + (5a - 3b + c)x + (3a + b + 2c) = 0$.
Therefore, (1) $2a + c = 0$ (2) $3a + b + 2c = 0$ (3) $-4a + 2b - c = 0$ (4) $5a - 3b + c = 0$ (5) $3a + b + 2c = 0$.

By addition of the five equations, we obtain (6) $9a + b + 5c = 0$.

From equation (1), $c = -2a$ or $-8a - 4c = 0$. Adding $-8a - 4c = 0$ to equation (6), we obtain $a + b + c = 0$.

COMMENT 1: Taking $a = 1$, $b = 1$, $c = -2$, verify the statement $P_1 + P_2 - 2P_3 = 0$.

COMMENT 2: Since $aP_1 + bP_2 + cP_3 = 0$ and, at least, one of a, b, and c is not equal to zero, and P_1, P_2, P_3 are said to be linearly dependent.

15-5 *If* $f_1 = 3x - y + 2z + w$,
   $f_2 = 2x + 3y - z + 2w$,
   $f_3 = 5x - 9y + 8z - w$, *find numerical values of* a, b, c *so that* $af_1 + bf_2 + cf_3 = 0$.

For $af_1 + bf_2 + cf_3 = 0$, $3ax + 2bx + 5cx = 0$, $-ay + 3by - 9cy = 0$, $2az - bz + 8cz = 0$, and $aw + 2bw - cw = 0$. (See Problem 15-4.)

$$3a + 2b + 5c = 0, x \neq 0$$
$$-a + 3b - 9c = 0, y \neq 0$$
$$2a - b + 8c = 0, z \neq 0$$
$$a + 2b - c = 0, w \neq 0$$

$\therefore a = -3c, b = 2c$

We can take $a = 3, b = -2$, and $c = -1$.

But we may also use any values for a, b, c that are proportional to $3, -2, -1$.

COMMENT: f_1, f_2, f_3 are linearly dependent.

15-6 *Find the common solutions of the set of equations*

$$x - 2xy + 2y = -1 \tag{I}$$
$$x - xy + y = 0. \tag{II}$$

$x = 1$ (equation II $-$ equation I) $\therefore xy - y = 1$ and $y = \dfrac{1}{x - 1}$.

Consequently, there are no finite solutions.

The geometric picture makes this clear (Fig. S15-6). Neither curve I $(x - 2xy + 2y = -1)$, nor curve II, $(x - xy + y = 0)$, crosses the line $x = 1$, indicating that the value $x = 1$ fails to satisfy either equation. However, as we approach arbitrarily close to $x = 1$, both on the right and on the left, the curves approach arbitrarily close to the line $x = 1$. Such a line is known as an asymptote.

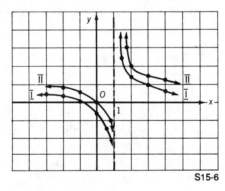

S15-6

15-7 *Solve the system* $3x + 4y + 5z = a,$
$\qquad\qquad\qquad 4x + 5y + 6z = b,$
$\qquad\qquad\qquad 5x + 6y + 7z = c,$

a, b, c *arbitrary real numbers, subject to the restriction* $x \geq 0,$ $y \geq 0, z > 0.$

For $a = b = c = 0$, there is no solution because of the restrictions $x \geq 0, y \geq 0, z > 0$.

Since the system determinant $D = \begin{vmatrix} 3 & 4 & 5 \\ 4 & 5 & 6 \\ 5 & 6 & 7 \end{vmatrix} = 0$,

there can be solutions only if $D_x = 0$, $D_y = 0$, and $D_z = 0$ where

$$D_x = \begin{vmatrix} a & 4 & 5 \\ b & 5 & 6 \\ c & 6 & 7 \end{vmatrix}, \quad D_y = \begin{vmatrix} 3 & a & 5 \\ 4 & b & 6 \\ 5 & c & 7 \end{vmatrix}, \quad \text{and} \quad D_z = \begin{vmatrix} 3 & 4 & a \\ 4 & 5 & b \\ 5 & 6 & c \end{vmatrix}.$$

(Cramer's Rule; Appendix VII)

Setting $D_x = 0$, then, we find $2b = a + c$ (setting $D_y = 0$ or $D_z = 0$ yields the same result). But $2b = a + c$ implies that a, b, c are in arithmetic sequence.

With a, b, c in arithmetic sequence, we have the one independent equation $4x + 5y + 6z = b$. However, to insure that $z > 0$, $b > 4x + 5y$. We then obtain $z = \frac{1}{6}(b - 4x - 5y)$ where $b - 4x - 5y > 0$. An infinite number of solutions is then obtained by assigning appropriate values to x and to y and solving for z.

ILLUSTRATION: Here are three solution sets just for the one choice of $b = 15$. (1) $x = 1$, $y = 1$, $z = 1$ (with $d = 3$) (2) $x = 1$, $y = 2$, $z = \frac{1}{6}$ $\left(\text{with } d = 3\frac{1}{6}\right)$ (3) $x = 2$, $y = 1$, $z = \frac{1}{3}$ $\left(\text{with } d = 3\frac{1}{3}\right)$ where d represents the common difference of the arithmetic sequence.

15-8 *For a class of* N *students,* $15 < N < 30$, *the following data were obtained from a test on which 65 or above is passing: the range of marks was from 30 to 90; the average for all was 66, the average for those passing was 71, and the average for those failing was 56. Based on a minor flaw in the wording of a problem, an upward adjustment of 5 points was made for all. Now the average mark of those passing became 79, and of those failing, 47. Find the number* N_0 *of students who passed originally, and the number* N_f *of those passing after adjustment, and* N.

$66N = 71N_0 + 56(N - N_0)$, and $71N = 79N_f + 47(N - N_f)$.

$$\therefore N_0 = \frac{2}{3}N, \quad N_f = \frac{3}{4}N, \quad \text{and} \quad N_f = \frac{9}{8}N_0.$$

$N_0 = \frac{8}{9} N_f$ so that N_0 is a multiple of 8 and N_f is a multiple of 9, since N_0 and N_f are integers.

However, since $N_0 = \frac{2}{3} N$ and $15 < N < 30$, $10 < N_0 < 20$, and, since $N_f = \frac{9}{8} N_0$, $11\frac{1}{4} < N_f < 22\frac{1}{2}$.

Since N_0 is a multiple of 8, and $10 < N_0 < 20$, $N_0 = 16$.

Since N_f is a multiple of 9, and $11\frac{1}{4} < N_f < 22\frac{1}{2}$, $N_f = 18$.

Since $N_0 = \frac{2}{3} N$ and $N_0 = 16$, $N = 24$.

16 Algebra and Geometry: Often the Twain Shall Meet

16-1 *Curve I is the set of points* (x, y) *such that* x = u + 1, y = −2u + 3, u *a real number. Curve II is the set of points* (x, y) *such that* x = −2v + 2, y = 4v + 1, v *a real number. Find the number of common points.*

Since $x = u + 1$, $u = x - 1$, and, since $y = -2u + 3$, then $u = \frac{3 - y}{2}$. Thus $x - 1 = \frac{3 - y}{2}$, or $2x + y = 5$. Therefore, curve I is a straight line. Since $x = -2v + 2$, $v = \frac{2 - x}{2}$, and, since $y = 4v + 1$, then $v = \frac{y - 1}{4}$. Thus $\frac{2 - x}{2} = \frac{y - 1}{4}$, or $2x + y = 5$. Therefore, curve II is a straight line coinciding with curve I. The number of common points is infinite.

16-2 *Let the altitudes of equilateral triangle* ABC *be* $\overline{AA_1}$, $\overline{BB_1}$, *and* $\overline{CC_1}$, *with intersection point* H. *Let p represent a counterclockwise rotation of the triangle in its plane through 120° about point* H. *Let q represent a similar rotation through 240°. Let r represent a rotation of the triangle through 180° about line* $\overleftrightarrow{AA_1}$. *And let* s, t *represent similar rotations about lines* $\overleftrightarrow{BB_1}$, $\overleftrightarrow{CC_1}$, *respectively.*

If we define p * r *to mean "first perform rotation* p *and then perform rotation* r," *find a simpler expression for* (q * r) * q; *that is, rotation* q *followed by rotation* r, *and this resulting rotation followed by rotation* q.

The rotation q followed by the rotation r is the equivalent of the single rotation s; that is, $q * r = s$ (Fig. S16-2). The rotation s followed by the rotation q is the equivalent of the single rotation t; that is, $s * q = t$. Therefore, $(q * r) * q = t$.

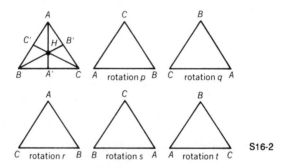

S16-2

16-3 *Figure S16-3 represents a transformation of the segment* $\overline{AB}$ *onto segment* $\overline{A'B'}$, *and of* $\overline{BC}$ *onto* $\overline{B'C'}$. *The points of* $\overline{AB}$ *go into points of* $\overline{A'B'}$ *by parallel projections (parallel to* $\overline{AA'}$). *The points of* $\overline{BC}$ *go into points of* $\overline{B'C'}$ *by projections through the fixed point* P.

The distances from the left vertical line $\overline{AM}$ *are zero for point* A, 3 *for point* B, 4 *for point* C, 5 *for point* B', *and* 2 *for point* A'(C'). *Designate the distances of the points on* $\overline{AC}$ *from* $\overline{AM}$ *as* x, *and the distances of their projections on* $\overline{A'B'}(\overline{C'B'})$ *from* $\overline{AM}$ *as* f. *Find the values of* r *and* s *of the transformation functions* f = rx + s **(a)** *for* $0 \le x \le 3$ **(b)** *for* $3 \le x \le 4$.

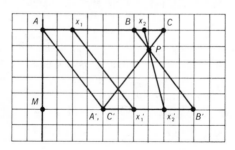

S16-3

The transformation function is $f = rx + s$ where f is the distance on $\overline{A'B'}$ from $\overline{AM}$ of the image point of a point on $\overline{AB}$ whose distance from $\overline{AM}$ is x. (a) For the point A, $x = 0$, and for the image point A', $f = 2$. Thus, $2 = 0 + s$ so that $s = 2$. For the point B, $x = 3$, and for the image point B', $f = 5$, so $5 = 3r + s$. Since $s = 2$, $5 = 3r + 2$ so that $r = 1$.

Therefore, the first transformation function is $f = x + 2$, $0 \leq x \leq 3$.

(b) For the point B, $x = 3$, and for the image point B', $f = 5$, so $5 = 3r + s$. For the point C, $x = 4$, and for the image point C', $f = 2$, so that $2 = 4r + s$. Solving the pair of equations $5 = 3r + s$ and $2 = 4r + s$, we find $r = -3$ and $s = 14$. Therefore, the second transformation function is $f = -3x + 14$, $3 \leq x \leq 4$.

16-4 *Given the three equations* (1) $7x - 12y = 42$ (2) $7x + 20y = 98$ (3) $21x + 12y = m$, *find the value(s) of* m *for which the three lines form a triangle of zero area.*

The triangle formed will have zero area when the lines represented by the equations are concurrent.

Lines (1) and (2) intersect in $\left(9, \dfrac{7}{4}\right)$. For line (3) to pass through this point, the coordinates must satisfy equation (3).

$$\therefore 21(9) + 12\left(\frac{7}{4}\right) = m, \quad m = 210$$

16-5 *Describe the graph of* $\sqrt{x^2 + y^2} = y$.

$x^2 + y^2 = y^2$, $x^2 = 0$, $x = 0$ (the y-axis). However, the given equation implies that $y \geq 0$, since $\sqrt{x^2 + y^2}$ is a non-negative number. Hence, the graph is that part of the y-axis such that $0 \leq y < \infty$, that is, the part on and above the x-axis.

16-6 *Transform* $x^2 - 3x - 5 = 0$ *into an equation of the form* $aX^2 + b = 0$ *where* a *and* b *are integers.*

Let $x = X + c$; then, by substitution,
$X^2 + 2cX + c^2 - 3X - 3c - 5 = 0$.
$X^2 + (2c - 3)X + c^2 - 3c - 5 = 0$
$\therefore 2c - 3 = 0$, and $c = \dfrac{3}{2}$, $c^2 - 3c - 5 = -\dfrac{29}{4}$.

Thus $X^2 - \frac{29}{4} = 0$, or $4X^2 - 29 = 0$, is the required equation.

16-7 *It is required to transform* $2x_1^2 - 4x_1x_2 + 3x_2^2$ *into an expression of the type* $a_1y_1^2 + a_2y_2^2$. *Using the transformation formulas* $y_1 = x_1 + cx_2$ *and* $y_2 = x_2$, *determine the values of* a_1 *and* a_2.

METHOD I: Substitute into the expression $a_1y_1^2 + a_2y_2^2$ the formulas $y_1 = x_1 + cx_2$ and $y_2 = x_2$ to obtain $a_1(x_1 + cx_2)^2 + a_2(x_2)^2$. Expanding this last expression, we obtain

$$a_1x_1^2 + 2a_1cx_1x_2 + a_1c^2x_2^2 + a_2x_2^2. \tag{I}$$

By the conditions of the problem, (I) is identically equal to $2x_1^2 - 4x_1x_2 + 3x_2^2$.

Therefore, $a_1 = 2$, $2a_1c = -4$ so that $c = -1$, and $a_1 + a_2 = 3$ so that $a_2 = 1$.

METHOD II: From the formula $y_1 = x_1 + cx_2$ we have $x_1 = y_1 - cx_2$, so that $x_1 = y_1 - cy_2$ since $y_2 = x_2$. Substituting these values into the expression $2x_1^2 - 4x_1x_2 + 3x_2^2$, we obtain, after simplification,

$$2y_1^2 - (4c + 4)y_1y_2 + (2c^2 + 3 + 4c)y_2^2. \tag{II}$$

Set (II) identically equal to $a_1y_1^2 + a_2y_2^2$. Therefore, $a_1 = 2$, $4c + 4 = 0$ so that $c = -1$, and $2c^2 + 3 + 4c = a_2$ so that $a_2 = 2 + 3 - 4 = 1$.

16-8 N.B. *and* S.B. *are, respectively, the north and south banks of a river with a uniform width of one mile. (See Fig. S16-8a.) Town* A *is 3 miles north of* N.B., *town* B *is 5 miles south of* S.B. *and 15 miles east of* A. *If crossing at the river banks is only at right angles to the banks, find the length of the shortest path from* A *to* B.

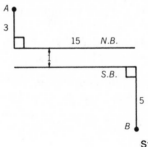

S16-8a

Consider the banks merged; then, obviously, the shortest path is the segment $\overline{AB}$. (See Fig. S16-8b.) $AB^2 = 15^2 + 8^2$, $AB = 17$. Since at the crossing point C there is a displacement of 1 mile, the shortest path is $17 + 1 = 18$ miles.

From the proportionality of the sides of similar triangles, we find that the crossing point is $5\frac{5}{8}$ miles east of A on $N.B.$

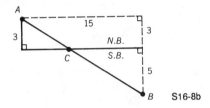

S16-8b

Challenge   *If the rate of land travel is uniformly 8 m.p.h., and the rowing rate on the river is $1\frac{2}{3}$ m.p.h. (in still water) with a west to east current of $1\frac{1}{3}$ m.p.h., find the shortest time it takes to go from A to B.*

The time required for land travel is $17 \div 8 = 2\frac{1}{8}$ hours. For the river crossing, the boat is pointed in the direction of segment $\overline{CD}$ whose length is $1\frac{2}{3}$ miles (Fig. S16-8c). Hence, the time required for the river crossing is $1\frac{2}{3} \div 1\frac{2}{3} = 1$ hour. The total time is, consequently, $2\frac{1}{8} + 1 = 3\frac{1}{8}$ hours.

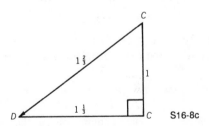

S16-8c

16-9 *Let the vertices of a triangle be $(0, 0)$, $(x, 0)$, and (hx, mx), m a positive constant and $0 \leq h < \infty$. Let a curve C be such that the y-coordinates of its points are numerically equal to the areas of the triangles for the values of h designated. Write the equation of the curve C.*

Let the segment determined by $(0, 0)$ and $(x, 0)$ be the base of the triangle; then the altitude to this base is mx. Therefore, $y = \frac{1}{2}x(mx) = \frac{1}{2}mx^2$. If x is restricted to the values $0 \le x < \infty$, the curve is the right half of the parabola $y = \frac{1}{2}mx^2$.

Challenge *Let the vertices of a triangle be $(0, 0)$, $(x, 0)$, and $\left(\frac{x}{2}, mx\right)$, m a positive constant. Let a curve C be such that the y-coordinates of its points are numerically equal to the perimeters of the triangles thus formed. Write the equation of curve C.*

The equation is $y = x(1 + \sqrt{1 + 4m^2})$, a straight line.

16-10 *Each member of the family of parabolas* $y = ax^2 + 2x + 3$ *has a maximum or a minimum point dependent upon the value of a. Find an equation of the locus of the maxima and minima for all possible values of a.*

METHOD I: Consider the general case $y = ax^2 + bx + c$. The turning point (maximum or minimum) is on the axis of symmetry of the parabola, whose equation is $x = -\frac{b}{2a}$, and, hence, the x-coordinate of the turning point is $-\frac{b}{2a}$. To find the y-coordinate of the turning point, substitute $-\frac{b}{2a}$ for x in $y = ax^2 + bx + c$. After simplification we find $y = -\frac{b^2}{4a} + c$. Since, however, $-\frac{b}{2a} = x$, we have $y = \frac{b}{2}\left(-\frac{b}{2a}\right) + c = \frac{b}{2}x + c$. Therefore, for the general case, the required locus equation is $y = \frac{b}{2}x + c$ (independent of a!).

In the given problem $b = 2$ and $c = 3$. Hence, the required equation for the given problem is $y = x + 3$.

METHOD II: In addition to the turning point with coordinates $\left(-\frac{b}{2a}, \frac{-b^2}{4a} + c\right)$, the required locus contains the point $(0, c)$. Therefore, for the general case, the required equation is $\dfrac{y - c}{x - 0} = \dfrac{\dfrac{-b^2}{4a} + c - c}{-\dfrac{b}{2a} - 0}$, which, after simplification, becomes $y = \frac{b}{2}x + c$. For the particular case where $b = 2$, $c = 3$, and $a = a$, we have $y = x + 3$.

16-11 *Let* A, B, *and* C *be three distinct points in a plane, such that* AB = X > 0, AC = 2AB, *and* AB + BC = AC + 2. *Find the values of* X *for which the three points may be the vertices of a triangle.*

$AB = X$. Since $AC = 2AB$, $AC = 2X$. Since $AB + BC = AC + 2$, $BC = X + 2$.

Since $AB + AC > BC$, $X + 2X > X + 2$, $\therefore X > 1$.

Since $AB + BC > AC$, $X + X + 2 > 2X$, $\therefore X$ is any (positive) number.

Since $AC + BC > AB$, $2X + X + 2 > X$, $\therefore X > -1$.

The intersection of these three sets of X-values is the set $X > 1$.

16-12 *The area of a given rectangle is* 450 *square inches. If the area remains the same when* h *inches are added to the width and* h *inches are subtracted from the length, find the new dimensions.*

Let L (inches) represent the original length, and let W (inches) represent the original width, with $L \geq W$. By the conditions of the problem $LW = 450$ and $(L - h)(W + h) = 450$.

Therefore, $LW = (L - h)(W + h)$ and so $LW = LW + h(L - W) - h^2$. Therefore, $h(L - W) - h^2 = 0$ or $h(L - W - h) = 0$. For $h \neq 0$, $L - W - h = 0$ so that $h = L - W$. Hence, the new length $L' = L - h = L - (L - W) = W$, and the new width $W' = W + h = W + (L - W) = L$.

COMMENT 1: When $h = 0$, $L = W$ since $h = L - W$. Then $L' = W = L$ and $W' = L = W$ so that there is no change in the dimensions.

COMMENT 2: When $h > 0$, $L > W$. Then $L' = W$ and $W' = L$, and the change is merely one in name, the original width becoming the new length and the original length becoming the new width. Why is this so?

COMMENT 3: An expression such as $LW = A$ where A is a constant, or, more generally, $xy = c$ where c is a constant, illustrates inverse variation in the variables x and y. To keep the product constant, a change in one variable needs to be offset by a "proper" change in the other. How do we determine the "proper" change?

COMMENT 4: Since $xy = c$, $y = \dfrac{c}{x}$. Let us, for example, change y to $\dfrac{4}{3}y$ so that $y' = \dfrac{4}{3}y$. Since, also, $x'y' = c$, then $x' = \dfrac{c}{y'}$, and, since $y' = \frac{4y}{3}$ and $y = \frac{c}{x}$, we have $x' = (c) \div \left(\frac{4}{3}\right)\left(\frac{c}{x}\right)$. There-fore, $x' = \frac{3x}{4}$. As verification we have $x'y' = \left(\frac{3x}{4}\right)\left(\frac{4y}{3}\right) = xy = c$. Note that the multiplier of y, $\frac{4}{3}$, is the reciprocal of the multiplier for x, $\frac{3}{4}$.

COMMENT 5: In the problem here posed, we have L changed to $L - h$ so that $L' = \left(\frac{L - h}{L}\right)L$, and W changed to $W + h$ so that $W' = \left(\frac{W + h}{W}\right)W$. Since the multiplying factors for L and W are reciprocals, their product is 1. Hence, $\frac{L - h}{L} \cdot \frac{W + h}{W} = 1$. When simplified, this equation becomes $h(L - W - h) = 0$, so that, as we have already found, $h = 0$ or $h = L - W$. These values for h lead, respectively, to the results $L' = L$ and $W' = W$, and $L' = W$ and $W' = L$. In Challenge 1 we illustrate further the importance of comment (4) above by introducing a significant change in the dimensions.

Challenge 1 *Solve the problem when* 12 *inches are added to the width, and* 10 *inches subtracted from the length.*

Let us increase the width by 12 inches and decrease the length by 10 inches. Then $LW = (L - 10)(W + 12)$. $\therefore LW = LW + 12L - 10W - 120 \therefore 12L - 10W = 120$, and so $W = \frac{12L - 120}{10}$.

Consequently, $LW = L\left(\frac{12L - 120}{10}\right) = 450$; $L^2 - 10L - 375 = 0$; $(L - 25)(L + 15) = 0$. $L = 25$ and $W = 18$. Therefore, $L' = L - 10 = 15$ and $W' = W + 12 = 30$.

We note that W was changed from 18 to 30, that is, $W' = \frac{30}{18}W = \frac{5}{3}W$, and that L was changed from 25 to 15, that is, $L' = \frac{15}{25}L = \frac{3}{5}L$. This illustrates the point made in comment (4) above that the multiplying factor for one of the variables, $\frac{5}{3}$, is the reciprocal of the multiplying factor for the second variable, $\frac{3}{5}$, when the product of the variables is to remain constant.

16-13 *Show that if the lengths of the sides of a triangle are represented by* a, b, *and* c, *a necessary and sufficient condition for the triangle to be equilateral is the equality* $a^2 + b^2 + c^2 = ab + bc + ca$.

That is, if the triangle is equilateral, then $a^2 + b^2 + c^2 = ab + bc + ca$, *and if* $a^2 + b^2 + c^2 = ab + bc + ca$, *then the triangle is equilateral.*

The condition is necessary. Given $\triangle ABC$ with $a = b = c$, prove $a^2 + b^2 + c^2 = ab + bc + ca$.

(1) $a = b$, so $ab = b^2$; $b = c$, so $bc = c^2$; $c = a$, so $ca = a^2$.

(2) $\therefore ab + bc + ca = a^2 + b^2 + c^2$

The condition is sufficient. Given $\triangle ABC$ with sides a, b, and c, and $a^2 + b^2 + c^2 = ab + bc + ca$, prove $a = b = c$.

(1) $a^2 + b^2 + c^2 = ab + bc + ca$

$\therefore c^2 - ca - cb + ab = -a^2 + 2ab - b^2$

$\therefore -(a - b)^2 = (c - a)(c - b)$

If $c > b > a$ or $c > a > b$, the left side of (1) is negative while the right side is positive, a contradiction.

(2) $-(b - c)^2 = (a - b)(a - c)$

If $a > b > c$ or $a > c > b$, the same contradiction appears.

(3) $-(c - a)^2 = (b - c)(b - a)$

If $b > c > a$ or $b > a > c$, the same contradiction appears.

$\therefore a = b = c$

16-14 *Select point* P *in side* $\overline{AB}$ *of triangle* ABC *so that* P *is between* A *and the midpoint of* $\overline{AB}$. *Draw the polygon (not convex)* $PP_1P_2P_3P_4P_5P_6$ *such that* $\overline{PP_1} \parallel \overline{AC}$, $\overline{P_1P_2} \parallel \overline{AB}$, $\overline{P_2P_3} \parallel \overline{CB}$, $\overline{P_3P_4} \parallel \overline{AC}$, $\overline{P_4P_5} \parallel \overline{AB}$, $P_5P_6 \parallel \overline{CB}$, *with* P_1, P_4 *in* $\overline{CB}$, P_2, P_5 *in* $\overline{AC}$, P_3, P_6 *in* $\overline{AB}$. *Show that point* P_6 *coincides with point* P.

Designate the length of $\overline{AP}$ by kc, where $0 < k < \frac{1}{2}$. Then $CP_1 = ka$, $CP_2 = kb$, $BP_3 = kc$, $BP_4 = ka$, $AP_5 = kb$, $AP_6 = kc$. But $AP = kc$. Therefore, P_6 coincides with P.

16-15 *Let* $P_1P_2P_3 \ldots P_nP_1$ *be a regular n-gon (that is, an n-sided polygon) inscribed in a circle with radius 1 and center at the origin, such that the coordinates of* P_1 *are* $(1, 0)$. *Let* $S = (P_1P_2)^2 + (P_1P_3)^2 + \cdots + (P_1P_n)^2$. *Find the ratio S:n.*

METHOD I: Designate the length of $\overline{P_1P_2}$ by d_1, the length of $\overline{P_1P_3}$ by d_2, $\ldots$, $\overline{P_1P_n}$ by d_{n-1}, and, in general, for any point P_i, the length of $\overline{P_1P_i}$ by d_{i-1}, where $i = 2, 3, \ldots, n$.

Using the Law of Cosines, we have

$$d_1{}^2 = 1^2 + 1^2 - 2 \cdot 1 \cdot 1 \cos \frac{2\pi}{n} = 2\left(1 - \cos \frac{2\pi}{n}\right)$$

$$d_2{}^2 = 1^2 + 1^2 - 2 \cdot 1 \cdot 1 \cos \frac{4\pi}{n} = 2\left(1 - \cos \frac{4\pi}{n}\right)$$

$$\vdots$$

$$d_{i-1}^2 = 1^2 + 1^2 - 2 \cdot 1 \cdot 1 \cos (i-1)\frac{2\pi}{n}$$

$$= 2\left(1 - \cos (i-1)\frac{2\pi}{n}\right)$$

$$\vdots$$

$$d_{\frac{n}{2}-(i-1)}^2 = 1^2 + 1^2 - 2 \cdot 1 \cdot 1 \cos \left[\frac{n}{2} - (i-1)\right]\frac{2\pi}{n}$$

$$= 2\left(1 + \cos (i-1)\frac{2\pi}{n}\right)$$

since $\left[\frac{n}{2} - (i-1)\right]\frac{2\pi}{n}$ is the supplement of $(i-1)\frac{2\pi}{n}$, where $\frac{n}{2} - (i-1) \geq 0$.

Therefore, $d_{i-1}^2 + d_{\frac{n}{2}-(i-1)}^2 = 2 \cdot 2$. When n is even, $P_{\frac{n}{2}+1}$ is one of the vertices, but when n is odd, $P_{\frac{n}{2}+1}$ is not one of the vertices. In either case $d_{\frac{n}{2}}$ is a diameter, since $d_{\frac{n}{2}}{}^2 = 2\left(1 - \cos \frac{n}{2} \cdot \frac{2\pi}{2}\right) = 2(1 - \cos \pi) = 2 \cdot 2$, so $d_{\frac{n}{2}} = 2$.

Therefore, $S = d_1{}^2 + d_2{}^2 + \cdots + d_{n-1}^2 = \sum\limits_{i=1}^{n-1} d_i{}^2 = \frac{n}{2} \cdot 2 \cdot 2 = 2n$. Hence, $S:n = 2n:n = 2:1$.

METHOD II: From right triangle P_1P_2A (see Fig. S16-15) we have

$$d_1{}^2 = \left(1 - \cos \frac{2\pi}{n}\right)^2 + \sin^2\left(\frac{2\pi}{n}\right)$$

$$d_1{}^2 = \left(1 - \cos \frac{2\pi}{n}\right)^2 - i^2 \sin^2\left(\frac{2\pi}{n}\right) \text{ (replacing 1 by } -i^2)$$

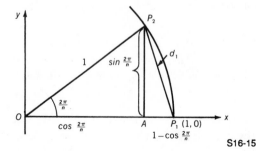

S16-15

Factoring the difference of two squares, we have

$$d_1{}^2 = \left(1 - \cos\frac{2\pi}{n} + i\sin\frac{2\pi}{n}\right)\left(1 - \cos\frac{2\pi}{n} - i\sin\frac{2\pi}{n}\right) \cdot$$

$$d_1{}^2 = \left[1 - \left(\cos\frac{-2\pi}{n} + i\sin\frac{-2\pi}{n}\right)\right]\left[1 - \left(\cos\frac{2\pi}{n} + i\sin\frac{2\pi}{n}\right)\right],$$

since $\cos\left(-\frac{2\pi}{n}\right) = \cos\frac{2\pi}{n}$ and $\sin\left(-\frac{2\pi}{n}\right) = -\sin\frac{2\pi}{n}$. We may now write $d_1{}^2 = (1 - w)(1 - w^{n-1})$ where w is the first imaginary root of the n n-th roots of 1 and w^{n-1} is the $(n - 1)$-th imaginary root. We, therefore, have $d_1{}^2 = 1 - w - w^{n-1} + w^n = 2 - w - w^{n-1}$ since $w^n = 1$.

In a similar manner we obtain

$$d_{n-1}^2 = (1 - w_{n-1})(1 - w) = 2 - w - w^{n-1},$$
and $d_2{}^2 = (1 - w^2)(1 - w^{n-2}) = 2 - w^2 - w^{n-2},$
and $d_{n-2}^2 = (1 - w_{n-2})(1 - w^2) = 2 - w^2 - w^{n-2},$
and so forth.

$$\therefore S = d_1{}^2 + d_2{}^2 + d_3{}^2 + \cdots$$
$$= 2n - 2(w + w^2 + w^3 + \cdots + w^{n-3} + w^{n-2} + w^{n-1})$$
$$\therefore S = 2n - 2\frac{w(1 - w^{n-1})}{1 - w}. \text{ (See Appendix VII.)}$$

Since, however, $w^n = 1$, $S = 2n$, and $S:n = 2:1$

---

17 Sequences and Series: Progression Procession

17-1 *Find the last two digits of* N $= 11^{10} - 1$.

$N = 11^{10} - 1 = (10 + 1)^{10} - 1 = 10^{10} + 10 \cdot 10^9 + \cdots + 10 \cdot 10 + 1 - 1$ (See Appendix VI.) Each of the terms $10^{10} + 10 \cdot 10^9 + \cdots + 10 \cdot 10$ ends in at least two zeros. Consequently, there is a common factor of 100 so that N ends in 00.

17-2 *Give a recursive definition of the sequence* $\left\{\dfrac{1}{4n}\right\}$, n *a natural number.*

Since the reciprocals of $\dfrac{1}{4n}$ are in arithmetic progression,

$$\frac{1}{f(n+1)} = \frac{1}{2}\left(\frac{1}{f(n)} + \frac{1}{f(n+2)}\right)$$

$$\therefore f(n+1) = \frac{2f(n)f(n+2)}{f(n)+f(n+2)}, \ f(1) = \frac{1}{4}, \ f(2) = \frac{1}{8}.$$

17-3 *A perfectly elastic ball is dropped from height* h *feet. It strikes a perfectly elastic surface* $\sqrt{\dfrac{h}{16}}$ *seconds later. It rebounds to a height* rh (*feet*), $0 < r < 1$, *to begin a similar bounce a second time, then a third time, and so forth. Find* (**a**) *the total distance* D (*feet*) *traveled and* (**b**) *the total time* T (*seconds*) *to travel* D *feet.*

$D = h + 2rh + 2r^2h + \cdots = \dfrac{2h}{1-r} - h = h\left(\dfrac{1+r}{1-r}\right)$ (See Appendix VII.)

$$T = \sqrt{\frac{h}{16}} + 2\sqrt{\frac{rh}{16}} + 2\sqrt{\frac{r^2h}{16}} + \cdots$$

$$= \frac{2\sqrt{\dfrac{h}{16}}}{1-\sqrt{r}} - \sqrt{\frac{h}{16}} = \frac{\sqrt{h}}{4}\cdot\frac{1+\sqrt{r}}{1-\sqrt{r}} \text{ (See Appendix VII.)}$$

17-4 *For the arithmetic sequence* a_1, a_2, ..., a_{16}, *it is known that* $a_7 + a_9 = a_{16}$. *Find each subsequence of three terms that forms a geometric sequence.*

$a_{16} = a_1 + 15d$, where d is the common difference, $a_7 = a_1 + 6d$ and $a_9 = a_1 + 8d$. Since $a_7 + a_9 = a_{16}$, $a_1 + 6d + a_1 + 8d = a_1 + 15d$. Therefore, $d = a_1$. Therefore, the i-th term of the sequence $a_i = a_1 + (i-1)d$, where $i = 1, 2, 3, \ldots,$ 16. Since $d = a_1$, $a_i = a_1 + ia_1 - a_1 = ia_1$ so that $a_1 = 1a_1$, $a_2 = 2a_1$, $a_3 = 3a_1, \ldots$.

Consequently, one subsequence forming a geometric sequence is a_1, a_2, a_4 (with a common ratio $r = 2$). A second subsequence is a_1, a_3, a_9 (with a common ratio $r = 3$). A third subsequence is a_1, a_4, a_{16} (with a common ratio $r = 4$). The fourth, fifth, and sixth subsequences are a_2, a_4, a_8 ($r = 2$); a_3, a_6, a_{12} ($r = 2$); a_4, a_8, a_{16} ($r = 2$).

For non-integer values of r such that $1 < r < 4$, we have only $r = \dfrac{3}{2}$ and $r = \dfrac{4}{3}$, so that additional sequences are a_4, a_6, a_9 and a_9, a_{12}, a_{16}.

17-5 *The sum of* n *terms of an arithmetic series is* 216. *The value of the first term is* n *and the value of the* n-*th term is* 2n. *Find the common difference,* d.

Since the sum of n terms of the given arithmetic series is 216, we have (see Appendix VII) $\frac{1}{2}n(n + 2n) = 216$, $n = 12$. (Why is $n = -12$ not acceptable?) $\therefore 2n = 24 = 12 + 11d$, $d = \frac{12}{11}$

17-6 *Find the sum of* n *terms of the arithmetic series whose first term is the sum of the first* n *natural numbers and whose common difference is* n.

Probably the difficult part of this problem is the seemingly confusing language!

Let T_n be the first term of the given series where $T_n = \frac{1}{2}n(n + 1)$. (See Appendix VII.) Then

$$S = T_n + (T_n + n) + (T_n + 2n) + \cdots + (T_n + (n - 1)n) = \frac{1}{2}n(2T_n + n^2 - n).$$
$$\therefore S = \frac{1}{2}n(n^2 + n + n^2 - n) = n^3$$

ILLUSTRATION: Let $n = 4$, then $T_n = \frac{1}{2}(4)(5) = 10$. The series is $10 + (10 + 4) + (10 + 8) + (10 + 12) = 64 = 4^3$.

Challenge *Prove that the sum of* n *terms of the arithmetic series, whose first term is the sum of the first* n *odd natural numbers, and whose common difference is* n, *is equal to the sum of* n *terms of the arithmetic series whose first term is the sum of the first* n *natural numbers and whose common difference is* 2n.

Let O_n be the first term of the series first described. Then $O_n = 1 + 3 + 5 + \cdots + (2n - 1) = n^2$. (See Appendix VII.)

$$S_1 = O_n + (O_n + n) + \cdots + (O_n + (n - 1)n)$$
$$= \frac{1}{2}n(2O_n + n^2 - n)$$
$$= \frac{1}{2}n(2n^2 + n^2 - n) = \frac{3}{2}n^3 - \frac{n^2}{2}$$

Let T_n be the first term of the second series. Then

$$S_2 = T_n + (T_n + 2n) + (T_n + 4n) + \cdots + (T_n + 2(n - 1)n)$$
$$= \frac{1}{2}n(2T_n + 2n^2 - 2n)$$
$$= \frac{1}{2}n(n^2 + n + 2n^2 - 2n) = \frac{3}{2}n^3 - \frac{n^2}{2}, \text{ so that } S_2 = S_1.$$

17-7 *In a given arithmetic sequence the* r*-th term is* s *and the* s*-th term is* r, r $\neq$ s. *Find the* (r + s)*-th term.*

Since the sequence is arithmetic, the r-th term is $a + (r - 1)d$, where a is the first term and d is the common difference. Therefore, $s = a + (r - 1)d$. In a similar manner we obtain

$$r = a + (s - 1)d. \tag{I}$$

By subtraction, $s - r = (r - s)d$. Dividing both sides of the equation by $r - s$, we have $d = -1$. Therefore, $a = r + s - 1$ (from I). Also, since the $(r + s)$-th term is $a + (r + s - 1)d$, we have the $(r + s)$-th term equal to $a + a(-1) = 0$ (substituting -1 for d and $r + s - 1$ for a).

Challenge *If* $S_n = \frac{1}{2}(r + s)(r + s - 1)$, *find* n.

Since $\frac{1}{2}(r + s)(r + s - 1) = \frac{1}{2}(r + s - 1)(r + s), n = r + s - 1$ or $r + s$.

Since the $(r + s)$-th term is zero, this is possible.

ILLUSTRATION: Let $r = 5$, $s = 9$. Then, since $d = -1$, $a = 13$. The first 14 terms of the sequence are 13, 12, 11, 10, 9, 8, 7, 6, 5, 4, 3, 2, 1, 0. The 5th term is 9, the 9th term is 5, the 14th term is zero, and the sum of thirteen terms equals the sum of fourteen terms.

17-8 *Define the triangular number* T_n *as* $T_n = \frac{1}{2}n(n + 1)$, *where* n = 0, 1, 2, ..., n, ..., *and the square number* S_n *as* $S_n = n^2$, *where* n = 0, 1, 2, ..., n, ... *Prove*
(*a*) $T_{n+1} = T_n + n + 1$ (*b*) $S_{n+1} = S_n + 2n + 1$ (*c*) $S_{n+1} = T_{n+1} + T_n$ (*d*) $S_n = 2T_n - n$. (*See Fig. S17-8.*)

(**a**) $T_{n+1} = \frac{1}{2}(n + 1)(n + 2) = \frac{1}{2}n(n + 1) + \frac{1}{2}(2n + 2)$

$= \frac{1}{2}n(n + 1) + n + 1 = T_n + n + 1$

(**b**) $S_{n+1} = (n + 1)^2 = n^2 + 2n + 1 = S_n + 2n + 1$

(**c**) $S_{n+1} = (n + 1)(n + 1) = \frac{1}{2}(n + 1)(n + n + 2)$

$= \frac{1}{2}(n + 1)(n + 2) + \frac{1}{2}n(n + 1) = T_{n+1} + T_n$

(**d**) $S_n = S_{n+1} - 2n - 1 = T_{n+1} + T_n - 2n - 1$
$= T_n + n + 1 + T_n - 2n - 1 = 2T_n - n$

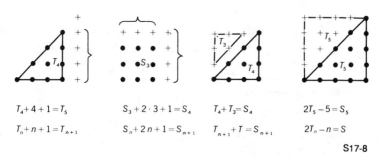

$T_4 + 4 + 1 = T_5$ $\qquad$ $S_3 + 2 \cdot 3 + 1 = S_4$ $\qquad$ $T_4 + T_3 = S_4$ $\qquad$ $2T_5 - 5 = S_5$

$T_n + n + 1 = T_{n+1}$ $\qquad$ $S_n + 2n + 1 = S_{n+1}$ $\qquad$ $T_{n+1} + T = S_{n+1}$ $\qquad$ $2T_n - n = S$

S17-8

17-9 *Beginning with the progression* a, ar, ar^2, ar^3, ..., ar^{n-1}, ...,
*form a new progression by taking for its terms the differences of
successive terms of the given progression, to wit,* ar $-$ a,
ar^2 $-$ ar, ... *Find the values of* a *and* r *for which the new progression is identical with the original.*

Since the progressions are the same, so are the $(n - 1)$-th terms.

$$ar^n - ar^{n-1} = ar^{n-1} \quad \therefore ar^n = 2ar^{n-1}$$

For $a \neq 0$, $r \neq 0$, $r = 2$. For $r = 2$, a is arbitrary.
ILLUSTRATION: From 3, 6, 12, 24, 48, ..., we obtain $6 - 3$,
$12 - 6$, $24 - 12$, $48 - 24$, ... ; that is 3, 6, 12, 24, ...

17-10 *The interior angles of a convex non-equiangular polygon of 9 sides
are in arithmetic progression. Find the least positive integer that
limits the upper value of the common difference between the
measures of the angles.*

We may represent the angle measures by a, $a + d$, ..., $a + 8d$.
$\therefore 9a + 36d = 7(180)$, or $a + 4d = 140$. Also, $a + 8d < 180$.
(Why?)
Therefore, $4d < 40$, $d < 10$; that is, the common difference is
less than $10°$.

17-11 *The division of* $\dfrac{s}{s + r}$, *where* r $\ll$ s, *that is, where* r *is very much
smaller in magnitude than* s, *is not exact, and is unending. If,
however, we agree to stop at a given point, the quotient is a polynomial in* $\dfrac{r}{s}$ *whose degree depends upon the stopping point. Find a
second-degree polynomial in* $\dfrac{r}{s}$ *best approximating the function*
$\dfrac{s}{s + r}$ *where* r $\ll$ s.

We may divide the numerator and the denominator of $\frac{s}{s+r}$ by s to obtain $\frac{1}{1+\frac{r}{s}}$. By actual division, or by binomial expansion, we find $\frac{1}{1+\frac{r}{s}} = 1 - \frac{r}{s} + \left(\frac{r}{s}\right)^2 - \left(\frac{r}{s}\right)^3 + \cdots$. Since $r \ll s$, $\left|\frac{r}{s}\right| < 1$. Therefore, we may truly say that $\frac{s}{s+r} = 1 - \frac{r}{s} + \left(\frac{r}{s}\right)^2 - \left(\frac{r}{s}\right)^3 + \cdots$. For a second-degree polynomial approximation to $\frac{s}{s+r}$, we therefore have $1 - \frac{r}{s} + \left(\frac{r}{s}\right)^2$.

ILLUSTRATION 1: Suppose $r = .1$ and $s = 10$. Using the approximating polynomial, we have $1 - \frac{.1}{10} + \left(\frac{.1}{10}\right)^2 = .9901$. By division we find that $\frac{s}{s+r} = \frac{10}{10 + .1} = \frac{10}{10.1} = .99009900 \ldots$. The approximate value agrees with the exact value in the fourth decimal place.

ILLUSTRATION 2: Suppose $r = -.1$ and $s = 10$. Using the approximating polynomial, we have $1 - \frac{-.1}{10} + \left(\frac{-.1}{10}\right)^2 = 1.0101$. Working directly with $\frac{s}{s+r}$, we have $\frac{10}{10 - .1} = \frac{10}{9.9} = 1.0101 \ldots$. Again there is agreement to four decimal places.

17-12 *When* $P_n(x) = 1 + x + x^2 + \cdots + x^n$ *is used to approximate the function* $P(x) = 1 + x + x^2 + \cdots + x^n + \cdots$ *when* $x = \frac{1}{4}$ *(see Problem* 17-11*), find the smallest integer* n *such that*

$$|P(x) - P_n(x)| < .001.$$

$P_n\left(\frac{1}{4}\right) = 1 + \frac{1}{4} + \frac{1}{4^2} + \cdots \frac{1}{4^n} = \frac{1}{3}\left(4 - \frac{1}{4^n}\right) = \frac{4}{3} - \frac{1}{3 \cdot 4^n}$ (see Appendix VII), while $P\left(\frac{1}{4}\right) = \frac{1}{1 - \frac{1}{4}} = \frac{4}{3}$.

When $x = \frac{1}{4}$, $|P(x) - P_n(x)| = \left|\frac{4}{3} - \left(\frac{4}{3} - \frac{1}{3 \cdot 4^n}\right)\right| = \frac{1}{3 \cdot 4^n} < \frac{1}{1000}$, so $n \geq 5$, when n is an integer.

Challenge 3 *Show that the values of* n *are large for those values of* x *that are toward the middle of the interval* $(-1, +1)$.

HINT FOR PROOF: First show that for any x in $-1 < x < 1$, $|P(x) - P_n(x)| = \left| \dfrac{1}{|x|^n(|x| - 1)} \right|$.

17-13 *Find the numerical value of* S *such that* S $= a_0 + a_1 + a_2 + \cdots + a_n + \cdots$ *where* $a_0 = 1$, $a_n = r^n$, *and* $a_{n+2} = a_n - a_{n+1}$.

$a_2 = a_0 - a_1$ $\therefore r^2 = 1 - r$, $r^2 + r - 1 = 0$, $r = \dfrac{-1 \pm \sqrt{5}}{2}$

Therefore, $S = \dfrac{1}{1 - \dfrac{-1 + \sqrt{5}}{2}} = \dfrac{3 + \sqrt{5}}{2}$, or

$S = \dfrac{1}{1 - \dfrac{-1 - \sqrt{5}}{2}} = \dfrac{3 - \sqrt{5}}{2}$.

Challenge   *Solve the problem with* $a_0 = 2$.

$r = -2$ or 1; but both must be rejected. Why?

17-14 *A group of men working together at the same rate can finish a job in 45 hours. However, the men report to work singly at equal intervals over a period of time. Once on the job, however, each man stays until the job is finished. If the first man works five times as many hours as the last man, find the number of hours the first man works.*

The number of man-hours required for the job is $h = 45n$ where n represents the number of men. Let x represent the number of hours the first man works. Then the second man works $(x - d)$ hours where d is the hour interval between the first and second man, and the n-th man works $x - (n - 1)d$ hours.
Let $S = x + (x - d) + (x - 2d) + \cdots + (x - (n - 1)d) = \frac{1}{2}n[2x - (n - 1)d]$. (See Appendix VII.)

$$\therefore \frac{1}{2}n[2x - (n - 1)d] = h = 45n \qquad (I)$$

Since $x = 5[x - (n - 1)d]$, $d = \dfrac{4x}{5(n - 1)}$. Substituting the value for d into (I), we have $\frac{1}{2}n\left(2x - \dfrac{4x}{5}\right) = 45n$. Therefore, $3x = 225$ and $x = 75$.

17-15 *A sequence of positive terms* A_1, A_2, $\ldots$, A_n, $\ldots$ *satisfies the recursive relation* $A_{n+1} = \dfrac{3(1 + A_n)}{3 + A_n}$. *For what values of* A_1 *is* *the sequence monotone decreasing* (i.e., $A_1 \geq A_2 \geq \cdots \geq A_n \geq \cdots$)?

For a monotone decreasing sequence, $A_1 \geq A_2 \geq \cdots \geq A_n \geq \cdots$.

Therefore, $A_2 = \dfrac{3 + 3A_1}{3 + A_1} \leq A_1$.

$3 + 3A_1 \leq 3A_1 + A_1{}^2; \sqrt{3} \leq A_1$

17-16 *If* $S = n^3 + (n + 1)^3 + (n + 2)^3 + \cdots + (2n)^3$, n *a positive integer, find* S *in closed form (that is, find a formula for* S*), given that* $1^3 + 2^3 + \cdots + n^3 = \dfrac{1}{4} n^2(n + 1)^2$.

Let $S_1 = 1^3 + 2^3 + 3^3 + \cdots + (2n)^3 = \dfrac{1}{4}(2n)^2(2n + 1)^2$
(using the given formula for 1, 2, ..., 2n).
Let $S_2 = 1^3 + 2^3 + 3^3 + \cdots + (n - 1)^3 = \dfrac{1}{4}(n - 1)^2(n)^2$
(using the given formula for 1, 2, ..., n − 1).

$\therefore S = S_1 - S_2 = \dfrac{1}{4}(2n)^2(2n + 1)^2 - \dfrac{1}{4}(n - 1)^2(n)^2$

$\quad = \dfrac{3}{4} n^2(n + 1)(5n + 1)$

ILLUSTRATION 1: Find S where $S = 3^3 + 4^3 + 5^3 + 6^3$.

$S = \dfrac{3}{4} \cdot 3^2 \cdot 4 \cdot 16 = 432$

ILLUSTRATION 2: Find S where $S = 50^3 + 51^3 + \cdots + 100^3$.

$S = \dfrac{3}{4}(50)^2(51)(251) = 24{,}001{,}875$

17-17 *If* $S(k) = 1 + 2 + 3 + \cdots + k$, *express* mn *in terms of* S(m), S(n), *and* S(m + n).

$S(m + n) = 1 + 2 + 3 + \cdots + (m + n)$
$\quad = \dfrac{1}{2}(m + n)(m + n + 1)$
$\quad = \dfrac{1}{2}(m^2 + 2mn + n^2 + m + n)$
$\quad = \dfrac{1}{2}(m^2 + m) + \dfrac{1}{2}(n^2 + n) + mn$

$S(m) = 1 + 2 + 3 + \cdots + m = \dfrac{1}{2} m(m + 1) = \dfrac{1}{2}(m^2 + m)$

Similarly, $S(n) = \dfrac{1}{2}(n^2 + n)$.

$S(m + n) = S(m) + S(n) + mn$
$mn = S(m + n) - S(m) - S(n)$

17-18 *Each* a_i *of the arithmetic sequence* a_0, a_1, 25, a_3, a_4 *is a positive integer. In the sequence there is a pair of consecutive terms whose squares differ by 399. Find the largest term of the sequence.*

METHOD I: Try $25^2 - a_1^2 = 399$. Then $a_1^2 = 226$. Reject, since a_1 is not an integer.

Try $a_3^2 - 25^2 = 399$. Then $a_3^2 = 1024 = 32^2$. $\therefore a_3 = 32$ and $d = 7$. Therefore, the largest term is $a_4 = 39$.

METHOD II: $a_0 = 25 - 2d$, $a_1 = 25 - d$, $a_3 = 25 + d$, $a_4 = 25 + 2d$.

Try $25^2 - (25 - d)^2 = 399$. Then $d^2 - 50d + 399 = 0$. The discriminant is 904, so that d is not an integer. Reject.

Try $(25 + d)^2 - 25^2 = 399$. Then $d^2 + 50d - 399 = 0 = (d - 7)(d + 57)$. Using the value $d = 7$, we obtain $a_3 = 32$ and $a_4 = 39$. With the value $d = -57$, the sequence is 139, 82, 25, -32, -89. With the condition in the problem that each a_i is a positive integer, this result is rejected. If the problem is changed so that each a_i is an integer, then the result 139 is acceptable.

17-19 *Let* $S_1 = 1 + cos^2 x + cos^4 x + \cdots$; *let* $S_2 = 1 + sin^2 x + sin^4 x + \cdots$; *let* $S_3 = 1 + sin^2 x \, cos^2 x + sin^4 x \, cos^4 x + \cdots$, *with* $0 < x < \frac{\pi}{2}$. *Show that* $S_1 + S_2 = S_1 S_2$, *and that* $S_1 + S_2 + S_3 = S_1 S_2 S_3$.

For $0 < x < \frac{\pi}{2}$, $0 < \cos x < 1$ $\therefore S_1 = \dfrac{1}{1 - \cos^2 x} = \dfrac{1}{\sin^2 x}$

(See Appendix VII.)

$0 < \sin x < 1$ $\therefore S_2 = \dfrac{1}{1 - \sin^2 x} = \dfrac{1}{\cos^2 x}$

$S_1 + S_2 = \dfrac{1}{\sin^2 x} + \dfrac{1}{\cos^2 x} = \dfrac{\cos^2 x + \sin^2 x}{\sin^2 x \cos^2 x} = \dfrac{1}{\sin^2 x \cos^2 x}$

$= S_1 S_2$ (since $\cos^2 x + \sin^2 x = 1$)

$S_3 = \dfrac{1}{1 - \sin^2 x \cos^2 x}$

$\therefore S_1 + S_2 + S_3 = \dfrac{1}{\sin^2 x} + \dfrac{1}{\cos^2 x} + \dfrac{1}{1 - \sin^2 x \cos^2 x}$

$= \dfrac{\cos^2 x(1 - \sin^2 x \cos^2 x) + \sin^2 x(1 - \sin^2 x \cos^2 x) + \sin^2 x \cos^2 x}{\sin^2 x \cos^2 x(1 - \sin^2 x \cos^2 x)}$

$= \dfrac{1}{\sin^2 x \cos^2 x(1 - \sin^2 x \cos^2 x)} = S_1 S_2 S_3$

17-20 *A square array of natural numbers is formed as shown. Find the sum of the elements in (**a**) the* j*-th column (**b**) the* i*-th row (**c**) the principal diagonal (upper left corner to lower right corner).*

1	2	3	.	.	.	n
n + 1	n + 2	.	.	.	.	2n
2n + 1	2n + 2	.	.	.	.	.
.	.	.	.	.	.	.
.	.	.	.	.	.	.
.	.	.	.	.	.	.
(n − 1)n + 1	(n − 1)n + 2	.	.	.	.	.

(**a**) In the *j*-th column the first element is *j* and the last element is $(n - 1)n + j$, with a common difference of *n*. Therefore, the sum $S_j = \frac{1}{2} n[j + (n - 1)n + j] = \frac{1}{2} n(n^2 - n + 2j)$.

(**b**) In the *i*-th row the first element is $(i - 1)n + 1$ and the last element is *ni*, with a common difference of 1. Therefore, the sum $S_i = \frac{1}{2} n[(i - 1)n + 1 + ni] = \frac{1}{2} n(2ni - n + 1)$.

(**c**) In the principal diagonal the first element is 1 and the last element is n^2, with a common difference of $n + 1$. Therefore, the sum $S_D = \frac{1}{2} n(1 + n^2) = \frac{1}{2} n(n^2 + 1)$.

17-21 *Let* $S = 2x + 2x^3 + 2x^5 + \cdots + 2x^{2k-1} + \cdots$, *where* $|x| < 1$, *be written as* $\frac{1}{P} - \frac{1}{Q}$. *Express* P *and* Q *as polynomials in* x *with integer coefficients.*

$S = 2x(1 + x^2 + x^4 + \cdots + x^{2k-2} + \cdots)$. Since $|x| < 1$, $S = 2x \frac{1}{1 - x^2} = \frac{2x}{1 - x^2}$.   (See Appendix VII.)

Let $\frac{2x}{1 - x^2} = \frac{A}{1 - x} + \frac{B}{1 + x}$ be an identity in *x*. Then $2x = A(1 + x) + B(1 - x)$. Letting $x = 1$, we find $A = 1$. Letting $x = -1$, we find $B = -1$. Therefore, $S = \frac{1}{1 - x} - \frac{1}{1 + x}$ so that $P = 1 - x$ and $Q = 1 + x$.

17-22 *Let* $I = \lim\limits_{n \to \infty} \dfrac{1^2 + 2^2 + \cdots + n^2}{n^3}$, *that is, the limiting value of the fraction as* n *increases without bound; find the value of* I.

Let $S = \dfrac{1}{n^3} (1^2 + 2^2 + \cdots + n^2)$

$\qquad = \dfrac{1}{n^3} \cdot \dfrac{1}{6} n(n + 1)(2n + 1).$ \qquad (See Appendix VII.)

$\qquad = \dfrac{2n^3 + 3n^2 + n}{6n^3} = \dfrac{2n^2 + 3n + 1}{6n^2}.$

Therefore, $S = \dfrac{1}{3} + \dfrac{1}{2n} + \dfrac{1}{6n^2}.$

As n increases without bound, both $\dfrac{1}{2n}$ and $\dfrac{1}{6n^2}$ approach arbitrarily close to zero. Therefore, the limiting value of S, which we labeled I, equals $\dfrac{1}{3}.$

In general, if $I = \lim\limits_{n \to \infty} \dfrac{1^r + 2^r + \cdots + n^r}{n^{r+1}}$, $I = \dfrac{1}{r + 1}.$ In this problem, $r = 2$ so that $I = \dfrac{1}{3}.$

A geometric interpretation is helpful (see Fig. S17-22). Consider the area K between the x-axis and the parabolic arc $y = x^2$, $0 \le x \le 1$. Partition $\overline{OA}$ into n equal segments and erect rectangles as shown. The sum of the areas of these rectangles approximates K, the approximation improving with increasing n. The heights of the rectangles are $\left(\dfrac{1}{n}\right)^2, \left(\dfrac{2}{n}\right)^2, \ldots, \left(\dfrac{n - 1}{n}\right)^2, \left(\dfrac{n}{n}\right)^2$, the widths are each equal to $\dfrac{1}{n}.$ Therefore, the sum K^* equals

$\dfrac{1}{n}\left(\dfrac{1}{n}\right)^2 + \dfrac{1}{n}\left(\dfrac{2}{n}\right)^2 + \cdots + \dfrac{1}{n}\left(\dfrac{n}{n}\right)^2 = \dfrac{1^2 + 2^2 + \cdots + n^2}{n^3}.$ As n increases without bound, $K^* \to K$. But K equals the area of rectangle $OABC$ minus the area of parabolic segment OBC.

$\therefore K = 1 - \dfrac{2}{3} = \dfrac{1}{3}.$ (We use a theorem of Archimedes which

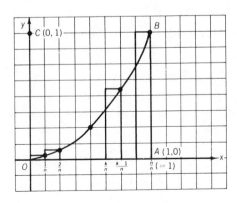

S17-22

states that the area of a parabolic segment is $\frac{2}{3}$ the area of the circumscribed rectangle.)

17-23 *Let* S $= \frac{2}{1 \cdot 2} + \frac{2}{2 \cdot 3} + \frac{2}{3 \cdot 4} + \cdots + \frac{2}{n(n + 1)}$. *Find a simple formula for* S.

Noting that $\frac{1}{n(n + 1)} = \frac{1}{n} - \frac{1}{n + 1}$, we may write S as

$$2\left[\frac{1}{1} - \frac{1}{2} + \frac{1}{2} - \frac{1}{3} + \frac{1}{3} - \frac{1}{4} + \cdots + \frac{1}{n - 1} - \frac{1}{n} + \frac{1}{n} - \frac{1}{n + 1}\right].$$

$\therefore S = 2\left(1 - \frac{1}{n + 1}\right) = \frac{2n}{n + 1}$, since all but the first and last terms are removed by "cancellation."

Challenge *Find the values of lim* S; *that is,* $\frac{2}{1 \cdot 2} + \frac{2}{2 \cdot 3} + \frac{2}{3 \cdot 4} + \cdots$.

$$\lim_{n \to \infty} S = \lim_{n \to \infty} 2\left(1 - \frac{1}{n + 1}\right) = 2$$

COMMENT: The terms of series S are the reciprocals of the triangular numbers $\frac{1}{2}n(n + 1)$. (See Appendix VII.)

17-24 *An endless series of rectangles is constructed on the curve* $\frac{1}{x}$, *each with width* 1 *and height* $\frac{1}{n} - \frac{1}{n + 1}$, n = 1, 2, 3, ... *Find the total area of the rectangles.*

METHOD I: Since the width of each rectangle is 1, the area is the same numerically as the height. The total area is, therefore,

$$\left(\frac{1}{1} - \frac{1}{2}\right) + \left(\frac{1}{2} - \frac{1}{3}\right) + \left(\frac{1}{3} - \frac{1}{4}\right) + \cdots + \left(\frac{1}{n - 1} - \frac{1}{n}\right) +$$

$$\left(\frac{1}{n} - \frac{1}{n + 1}\right) + \cdots = \lim_{n \to \infty}\left(1 - \frac{1}{n + 1}\right) = 1.$$

METHOD II: In Fig. S17-24, R_1 represents the rectangle with width 1 and height $\frac{1}{1} - \frac{1}{2}$, R_2 represents the rectangle with width 1 and height $\frac{1}{2} - \frac{1}{3}$, and so on.

Let the rectangles be translated to the left to come in contact with the y-axis, as shown by the dotted lines. Then $R_1 + R_2 + \cdots$ now occupy the rectangular region whose vertices are (0, 0), (1, 0), (1, 1), and (0, 1). The area of this region is $1 \times 1 = 1$. Therefore, the total area of the rectangles as described in the problem is 1.

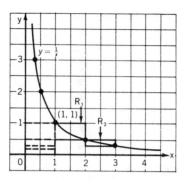

S17-24

COMMENT: An alternative way of stating this problem is, find S where $S = \frac{1}{2} + \frac{1}{6} + \frac{1}{12} + \cdots + \frac{1}{n(n+1)} + \cdots$ (See Problem 17-23.)

17-25 *Let* $S_n = \frac{1}{1 \cdot 4} + \frac{1}{4 \cdot 7} + \cdots + \frac{1}{(3n-2)(3n+1)}$ *where* n = 1, 2, ... *Find a simple formula for* S_n *in terms of* n.

We note that $u_n = \frac{1}{(3n-2)(3n+1)} = \frac{1}{3}\left(\frac{1}{3n-2} - \frac{1}{3n+1}\right)$.
Then
$$S_n = u_n + u_{n-1} + \cdots + u_2 + u_1$$
$$= \frac{1}{3}\left(\frac{1}{3n-2} - \frac{1}{3n+1}\right) + \frac{1}{3}\left(\frac{1}{3n-5} - \frac{1}{3n-2}\right) + \cdots$$
$$+ \frac{1}{3}\left(\frac{1}{4} - \frac{1}{7}\right) + \frac{1}{3}\left(\frac{1}{1} - \frac{1}{4}\right)$$
$$= \frac{1}{3}\left(1 - \frac{1}{3n+1}\right) = \frac{n}{3n+1}.$$
COMMENT: The limiting value of S_n as n increases without bound is $\frac{1}{3}$.

17-26 *Let* S = $a_1 + a_2 + \cdots + a_{n-1} + a_n$ *be a geometric series with common ratio* r, r $\neq$ 0, r $\neq$ 1. *Let* T = $b_1 + b_2 + \cdots + b_{n-1}$ *be the series such that* b_j *is the arithmetic mean (average) of* a_j *and* a_{j+1}, j = 1, 2, 3, ..., n. *Express* T *in terms of* a_1, a_n, *and* r.

As defined in the problem, $b_1 = \frac{1}{2}(a_1 + a_2)$, $b_2 = \frac{1}{2}(a_2 + a_3)$, and so on.

Since $a_1, a_2, a_3, \ldots$ is a geometric sequence with common ratio r, $a_2 = a_1r$, $a_3 = a_1r^2$, and so on.

Therefore, $b_1 = \frac{1}{2}a_1(1 + r)$, $b_2 = \frac{1}{2}a_1r(1 + r)$, $b_3 = \frac{1}{2}a_1r^2(1 + r)$, ..., $b_{n-1} = \frac{1}{2}a_1r^{n-2}(1 + r)$. Consequently, $T = \left(\frac{1 + r}{2}\right)(a_1)(1 + r + \cdots + r^{n-2}) = \left(\frac{1 + r}{2}\right)(a_1)\frac{1 - r^{n-1}}{1 - r}$.

(See Appendix VII.) $T = \frac{1 + r}{2(1 - r)}(a_1 - a_1r^{n-1}) = \frac{(1 + r)(a_1 - a_n)}{2(1 - r)}$ since $a_n = a_1r^{n-1}$.

17-27 *The sum of a number and its reciprocal is 1. Find the sum of the n-th power of the number and the n-th power of the reciprocal.*

METHOD I: We have

$x + \dfrac{1}{x} = 1$

$x^2 + \dfrac{1}{x^2} = \left(x + \dfrac{1}{x}\right)^2 - 2 = 1 - 2 = -1$

$x^3 + \dfrac{1}{x^3} = \left(x + \dfrac{1}{x}\right)\left(x^2 + \dfrac{1}{x^2}\right) - \left(x + \dfrac{1}{x}\right) = (1)(-1) - 1 = -2$

$x^4 + \dfrac{1}{x^4} = \left(x^2 + \dfrac{1}{x^2}\right)\left(x^2 + \dfrac{1}{x^2}\right) - 2 = (-1)(-1) - 2 = -1$

$x^5 + \dfrac{1}{x^5} = \left(x^3 + \dfrac{1}{x^3}\right)\left(x^2 + \dfrac{1}{x^2}\right) - \left(x + \dfrac{1}{x}\right) = (-2)(-1) - 1 = 1$

$x^6 + \dfrac{1}{x^6} = \left(x^3 + \dfrac{1}{x^3}\right)\left(x^3 + \dfrac{1}{x^3}\right) - 2 = (-2)(-2) - 2 = 2$

It will be seen shortly that it was not necessary to develop the values for $n = 5$ and $n = 3$ and even for $n = 6$.

We now show by mathematical induction (see Appendix VII) that $x^n + \dfrac{1}{x^n} = 2$ for all n of the form $6k$, $k = 0, 1, 2, \ldots$. For $k = 0$ we have $x^0 + \dfrac{1}{x^0} = 1 + 1 = 2$. For $k = 1$ we have $x^6 + \dfrac{1}{x^6} = 2$, as shown above.

Assume that $x^{6k} + \dfrac{1}{x^{6k}} = 2$, for an arbitrary integer value of $k \geq 1$. Then $x^{6(k+1)} + \dfrac{1}{x^{6(k+1)}} = \left(x^{6k} + \dfrac{1}{x^{6k}}\right)\left(x^6 + \dfrac{1}{x^6}\right) - \left(x^{6(k-1)} + \dfrac{1}{x^{6(k-1)}}\right) = 2 \cdot 2 - 2 = 2$, since $x^{6k} + \dfrac{1}{x^{6k}} = 2$ and $x^{6(k-1)} + \dfrac{1}{x^{6(k-1)}} = 2$ by assumption and $x^6 + \dfrac{1}{x^6} = 2$ by verification.

For $6k + 1$ and $6k + 5$, we may use the single form $6k \pm 1$ since $5 = 6 - 1$ and $x + \frac{1}{x} = x^{-1} + \frac{1}{x^{-1}}$. Since $x^{6k\pm1} +$

$\frac{1}{x^{6k\pm1}} = \left(x^{6k} + \frac{1}{x^{6k}}\right)\left(x^{\pm1} + \frac{1}{x^{\pm1}}\right) - \left(x^{6k\pm1} + \frac{1}{x^{6k\pm1}}\right)$, we have

$2\left(x^{6k\pm1} + \frac{1}{x^{6k\pm1}}\right) = \left(x^{6k} + \frac{1}{x^{6k}}\right)\left(x^{\pm1} + \frac{1}{x^{\pm1}}\right) = (2)(1)$. Therefore, $x^{6k\pm1} + \frac{1}{x^{6k\pm1}} = 1$.

For $6k + 2$ and $6k + 4$ we use the single form $6k \pm 2$, because $6k + 4 = 6(k + 1) - 6 + 4 = 6k - 2$. Since $x^{6k\pm2} +$

$\frac{1}{x^{6k\pm2}} = \left(x^{6k} + \frac{1}{x^{6k}}\right)\left(x^{\pm2} + \frac{1}{x^{\pm2}}\right) - \left(x^{6k\pm2} + \frac{1}{x^{6k\pm2}}\right)$, it follows

that $2\left(x^{6k\pm2} + \frac{1}{x^{6k\pm2}}\right) = \left(x^{6k} + \frac{1}{x^{6k}}\right)\left(x^{\pm2} + \frac{1}{x^{\pm2}}\right) = 2(-1)$ so

that $x^{6k\pm2} + \frac{1}{x^{6k\pm2}} = -1$.

In a similar manner, we have $x^{6k\pm3} + \frac{1}{x^{6k\pm3}} = -2$.

In summary, $x^n + \frac{1}{x^n} = $
$\quad$ 2 when $n = 6k$
$\quad$ 1 when $n = 6k \pm 1$
$\quad$ -1 when $n = 6k \pm 2$
$\quad$ -2 when $n = 6k \pm 3$.

METHOD II: Since $x + \frac{1}{x} = 1$, $x^2 - x + 1 = 0$, with roots

$$r_1 = \frac{1 + i\sqrt{3}}{2} = \frac{1}{r^2}, \text{ and } r_2 = \frac{1 - i\sqrt{3}}{2} = \frac{1}{r_1}.$$

Consider the equation $x^3 - 1 = 0$, with roots 1, w, w^2, where

$w = \frac{-1 + i\sqrt{3}}{2} = \frac{1}{w^2}$, $w^2 = \frac{-1 - i\sqrt{3}}{2} = \frac{1}{w}$.

Since $w^3 = 1$, $w^{3k} = 1$, and $1 + w + w^2 = 0$, and

$r_1 = \frac{1}{r_2} = -w^2$, and $r_2 = \frac{1}{r_1} = -w$.

Therefore, $x + \frac{1}{x} = -w^2 - w = 1$,

$$x^2 + \frac{1}{x^2} = w^4 + w^2 = w + w^2 = -1,$$

$$x^3 + \frac{1}{x^3} = -w^6 - w^3 = -1 - 1 = -2,$$

$$x^4 + \frac{1}{x^4} = w^8 + w^4 = w^2 + w = -1,$$

$$x^5 + \frac{1}{x^5} = -w^{10} - w^5 = -w - w^2 = 1,$$

$$x^6 + \frac{1}{x^6} = w^{12} + w^6 = 1 + 1 = 2.$$

Hence, $x^{6k} + \dfrac{1}{x^{6k}} = w^{12k} + w^{6k} = 1 + 1 = 2,$

$$x^{6k \pm 1} + \frac{1}{x^{6k \pm 1}} = -w^{\pm 2} - w^{\pm 1} = 1,$$

$$x^{6k \pm 2} + \frac{1}{x^{6k \pm 2}} = w^{\pm 4} + w^{\pm 2} = w^{\pm 1} + w^{\pm 2} = -1,$$

$$x^{6k \pm 3} + \frac{1}{x^{6k \pm 3}} = -w^{\pm 6} - w^{\pm 3} = -2.$$

METHOD III: Using trigonometry, we have $x + \dfrac{1}{x} = 1$, $x^2 - x + 1 = 0$, $r_1 = \dfrac{1 + i\sqrt{3}}{2} = \cos 60° + i \sin 60°$, $r_2 = \dfrac{1 - i\sqrt{3}}{2} = \cos 60° - i \sin 60°$, $r_1 = \dfrac{1}{r_2}$, $r_2 = \dfrac{1}{r_1}$.

Therefore, $x + \dfrac{1}{x} = (\cos 60° + i \sin 60°) + (\cos 60° - i \sin 60°)$

$$= \frac{1}{2} + \frac{i\sqrt{3}}{2} + \frac{1}{2} - \frac{i\sqrt{3}}{2} = 1.$$

Since $(\cos 60° \pm i \sin 60°)^n = \cos 60n° \pm i \sin 60n°$ (De Moivre Theorem), we conclude as follows.

$x^2 + \dfrac{1}{x^2} = (\cos 120° + i \sin 120°) + (\cos 120° - i \sin 120°) =$

$$-\frac{1}{2} + \frac{i\sqrt{3}}{2} - \frac{1}{2} - \frac{i\sqrt{3}}{2} = -1$$

$x^3 + \dfrac{1}{x^3} = (\cos 180° + i \sin 180°) + (\cos 180° - i \sin 180°) =$

$$-1 + 0 - 1 - 0 = -2$$

$x^4 + \dfrac{1}{x^4} = (\cos 240° + i \sin 240°) + (\cos 240° - i \sin 240°) =$

$$-\frac{1}{2} - \frac{i\sqrt{3}}{2} - \frac{1}{2} + \frac{i\sqrt{3}}{2} = -1$$

$x^5 + \dfrac{1}{x^5} = (\cos 300° + i \sin 300°) + (\cos 300° - i \sin 300°) =$

$$\frac{1}{2} - \frac{i\sqrt{3}}{2} + \frac{1}{2} + \frac{i\sqrt{3}}{2} = 1$$

$x^6 + \dfrac{1}{x^6} = (\cos 360° + i \sin 360°) + (\cos 360° - i \sin 360°) =$

$$1 + 0 + 1 - 0 = 2$$

$$x^{6k} + \frac{1}{x^{6k}} = (\cos 360k° + i \sin 360k°) +$$

$$(\cos 360k° - i \sin 360k°) =$$

$$1 + 0 + 1 - 0 = 2, \text{ and so on.}$$

17-28 *Alpha travels uniformly 20 miles a day. Beta, starting from the same point three days later to overtake Alpha, travels at a uniform rate of 15 miles the first day, at a uniform rate of 19 miles the second day, and so forth in arithmetic progression. If* n *represents the number of days Alpha has traveled when Beta overtakes him, find* n *(not necessarily an integer).*

Using a routine procedure, we may write

$$20(n + 3) = 15 + 19 + \cdots + [15 + (n - 1)4]$$
$$= \frac{1}{2}n[30 + (n - 1)4]$$

(See Appendix VII.) Then $2n^2 - 7n - 60 = 0 = (2n - 15) \times (n + 4)$. Therefore, $n = 7\frac{1}{2}$ and $n + 3 = 10\frac{1}{2}$.

Checking, we find Alpha's distance is 210 miles and Beta's distance is $210\frac{1}{2}$ miles. What is wrong?

We misused the formula $\frac{1}{2}n[2a + (n - 1)d]$; the formula is valid only for n a positive integer, as proved by mathematical induction. (See Appendix VII.)

After 10 days Alpha has traveled 200 miles while Beta has traveled 189 miles. Sometime during the eleventh day Alpha and Beta meet. Let x be the number of hours on the eleventh day when the meeting occurs. Then $\frac{x}{24} \cdot 43 = 11 + \frac{x}{24}(20)$, $23x = 11 \cdot 24$, $x = \frac{11}{23} \cdot 24$. Hence, $n = 10\frac{11}{23}$ days.

CHECK: In $10\frac{11}{23}$ days Alpha has traveled $200 + \frac{220}{23}$ miles while Beta has traveled $189 + \frac{11}{23} \cdot 43 = 189 + 11 + \frac{220}{23} = 200 + \frac{220}{23}$ miles.

Let us view the problem geometrically (Fig. S17-28). Alpha's distance is represented by the area of the series of rectangles $60 + 7 \cdot 20 + 20t$. Beta's distance is represented by the area of the rectangles $15 + 19 + 23 + \cdots + 39 + 43t$.

Below $\overline{RS}$ (see diagram) the excess of the Alpha area over the

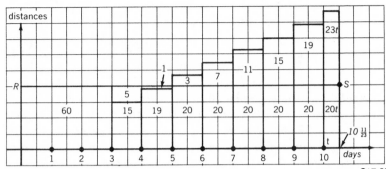

S17-28

Beta area is $60 + 5 + 1$. Above $\overline{RS}$ the excess of the Beta area over the Alpha area is $3 + 7 + 11 + 15 + 19 + 23t$.

$$\therefore 66 = 55 + 23t, \quad t = \frac{11}{23}$$

17-29 *Find a closed-form expression for* S_n, *where* $S_n = 1 \cdot 2 + 2 \cdot 2^2 + 3 \cdot 2^3 + \cdots + n \cdot 2^n$; *that is, find a simple formula for S.*

$$S_n = 1 \cdot 2 + 2 \cdot 2^2 + 3 \cdot 2^3 + \cdots + (n - 1)2^{n-1} + n \cdot 2^n$$
$$2S_n = 1 \cdot 2^2 + 2 \cdot 2^3 + \cdots + (n - 2)2^{n-1} + (n - 1)2^n + n \cdot 2^{n+1}$$

By subtraction, $S_n = n \cdot 2^{n+1} - (2 + 2^2 + 2^3 + \cdots + 2^n)$
$$= n \cdot 2^{n+1} - \frac{2(2^n - 1)}{2 - 1} = 2 + (n - 1)2^{n+1}$$

(See Appendix VII.)

ILLUSTRATION: For $n = 4$, $S_4 = 2 + 3 \cdot 2^5 = 98 = 2 + 8 + 24 + 64$.

17-30 *Show that* $\sum\limits_{r=1}^{\infty} \frac{1}{r^2} < 2$ *where* $\sum\limits_{r=1}^{\infty} \frac{1}{r^2} = \frac{1}{1^2} + \frac{1}{2^2} + \cdots + \frac{1}{r^2} + \cdots$.

First we show that $\sum\limits_{r=1}^{\infty} \frac{1}{r(r + 1)} = 1$, and that $\sum\limits_{r=2}^{\infty} \frac{1}{r(r - 1)} = 1$.

Since $\frac{1}{r(r + 1)} = \frac{1}{r} - \frac{1}{r + 1}$ (see Problem 17-23),

$$\sum_{r=1}^{\infty} \frac{1}{r(r + 1)} = \left(\frac{1}{1} - \frac{1}{2}\right) + \left(\frac{1}{2} - \frac{1}{3}\right) + \left(\frac{1}{3} - \frac{1}{4}\right) + \cdots = 1.$$

Since $\frac{1}{r(r - 1)} = \frac{1}{r - 1} - \frac{1}{r}$,

$$\sum_{r=2}^{\infty} \frac{1}{r(r - 1)} = \left(\frac{1}{1} - \frac{1}{2}\right) + \left(\frac{1}{2} - \frac{1}{3}\right) + \left(\frac{1}{3} - \frac{1}{4}\right) + \cdots = 1.$$

Since $\quad r(r-1) < r^2 < r(r+1), \quad \dfrac{1}{r(r+1)} < \dfrac{1}{r^2} < \dfrac{1}{r(r-1)}$.

Therefore,

$$\sum_{r=1}^{\infty} \frac{1}{r(r+1)} < \sum_{r=1}^{\infty} \frac{1}{r^2} \text{ and } \sum_{r=2}^{\infty} \frac{1}{r^2} < \sum_{r=2}^{\infty} \frac{1}{r(r-1)}.$$

$$1 + \sum_{r=2}^{\infty} \frac{1}{r^2} = \sum_{r=1}^{\infty} \frac{1}{r^2} < 1 + \sum_{r=2}^{\infty} \frac{1}{r(r-1)}$$

$$\sum_{r=1}^{\infty} \frac{1}{r(r+1)} < \sum_{r=1}^{\infty} \frac{1}{r^2} < 1 + \sum_{r=2}^{\infty} \frac{1}{r(r-1)}$$

Thus, $1 < \displaystyle\sum_{r=1}^{\infty} < \dfrac{1}{r^2} < 1 + 1 = 2$.

17-31 *Express* S_n *in terms of* n, *where* $S_n = 1 \cdot 1! + 2 \cdot 2! + 3 \cdot 3! + \cdots + n \cdot n!$.

METHOD I: $S_n = 1 \cdot 1! + 2 \cdot 2! + \cdots + n \cdot n! = 1 \cdot 1! + 2 \cdot 2! + \cdots + n \cdot n! + n! - n!$

But $n \cdot n! + n! = (n+1)!$. $\therefore S_n = 1 \cdot 1! + 2 \cdot 2! + \cdots + (n-1)(n-1)! - n! + (n+1)!$.

But $(n-1)(n-1)! - n! = -(n-1)!$ $\therefore S_n = 1 \cdot 1! + 2 \cdot 2! + \cdots + (n-2)(n-2)! - (n-1)! + (n+1)!$.

Continue in this manner to obtain $S_n = 1 \cdot 1! + 2 \cdot 2! - 3! + (n+1)!$.

Therefore, $S_n = (n+1)! - 1$.

METHOD II: Using mathematical induction, we have

$S_1 = 1 \cdot 1! = 1 = 2! - 1$ and $S_2 = 1 \cdot 1! + 2 \cdot 2! = 5 = 3! - 1$.

Assume that $S_k = (k+1)! - 1$; then $(k+1)(k+1)! + S_k = S_{k+1} = (k+1)! - 1 + (k+1)(k+1)!$.

Therefore, $S_{k+1} = (k+1)!(k+2) - 1 = (k+2)! - 1$.

18 Logarithms: A Power Play

18-1 *Find the real values of* x *such that* x $log_2 3 = log_{10} 3$.

Since $x \log_2 3 = \log_{10} 3$, $x = \dfrac{\log_{10} 3}{\log_2 3}$. Let $\log_2 3 = a$, then $2^a = 3$ (I).

Taking logarithms of both sides of (I) to the base 10, we have $a \log_{10} 2 = \log_{10} 3$.

Therefore, $x = \dfrac{a \log_{10} 2}{a} = \log_{10} 2$. Hence, $x \approx .3010$.

Challenge *Find the real values of* y *such that* y $log_{10} 3 = log_2 3$.

METHOD I: (Quickie Solution) Since y is the reciprocal of x, $y \approx \dfrac{1}{.3010}$.

METHOD II: (Formal Solution) Follow the pattern used in Problem 18-1.

18-2 *Find the real values of* x *for which* (**a**) F *is real* (**b**) F *is positive*, *where* F = $log_a \dfrac{2x + 4}{3x}$, a > 0, a ≠ 1.

(**a**) F is real when $0 < \dfrac{2x + 4}{3x}$ so that either $2x + 4 > 0$ and $3x > 0$ or $2x + 4 < 0$ and $3x < 0$. Therefore, F is real for x in the union of the sets $x > 0$ and $x < -2$.

(**b**) F is positive when $1 < \dfrac{2x + 4}{3x}$ so that $x < 4$. However, since F is not real for $-2 \le x \le 0$, F is positive only for the set $0 < x < 4$.

18-3 *If* f *is a function of* x *only and* g *is a function of* y *only, determine* f *and* g *such that* $log\, f + log\, g = log(1 + z)$ *where* z = x + xy + y.

$\log (1 + z) = \log (1 + x + xy + y)$
$= \log ((1 + x)(1 + y)) = \log f + \log g = \log fg$
$\therefore f = 1 + x$ and $g = 1 + y$.

18-4 *If* $(ax)^{log\, a} = (bx)^{log\, b}$, a, b *positive*, a ≠ b, a ≠ 1, b ≠ 1, *and the logarithmic base is the same throughout, express* x *in terms of* a *and* b.

Taking logarithms of both sides of the equation to the same base,

$$\log a \log ax = \log b \log bx$$
$$\log a(\log a + \log x) = \log b(\log b + \log x)$$
$$(\log a)^2 + \log a \log x = (\log b)^2 + \log b \log x$$
$$\log x(\log a - \log b) = (\log b)^2 - (\log a)^2$$
$$= (\log b + \log a)(\log b - \log a)$$
$$\log x = -(\log b + \log a)$$
$$= -\log ab$$
$$= \log (ab)^{-1}$$

Therefore, $\qquad x = (ab)^{-1}.$

VERIFICATION: $\left(a \cdot \dfrac{1}{ab}\right)^{\log a} \overset{?}{=} \left(b \cdot \dfrac{1}{ab}\right)^{\log b}$ or $\left(\dfrac{1}{b}\right)^{\log a} \overset{?}{=} \left(\dfrac{1}{a}\right)^{\log b}$.

Let $\log_c a = u$ so that $a = c^u$, and let $\log_c b = v$ so that $b = c^v$. $\left(\dfrac{1}{b}\right)^{\log_c a} = \left(\dfrac{1}{c^v}\right)^u = \dfrac{1}{c^{uv}}$ and $\left(\dfrac{1}{a}\right)^{\log_c b} = \left(\dfrac{1}{c^u}\right)^v = \dfrac{1}{c^{uv}}$, with reversible steps, or, taking logarithms of both sides of the tested equality $\left(\dfrac{1}{b}\right)^{\log a} \overset{?}{=} \left(\dfrac{1}{a}\right)^{\log b}$, we have $\log a(\log 1 - \log b) \overset{?}{=} \log b(\log 1 - \log a)$. Since $\log 1 = 0$, we have $-\log a \log b = -\log a \log b$ with reversible steps.

18-5 *Find a simple formula for* $S_n = \dfrac{1}{\log_2 N} + \dfrac{1}{\log_3 N} + \cdots + \dfrac{1}{\log_{25} N}$. $N > 1$.

Since $\log_a b = \dfrac{1}{\log_b a}$, then $S_n = \log_N 2 + \log_N 3 + \cdots + \log_N 25 = \log_N 25!$

Challenge *Find a simple formula for* $T_n = \dfrac{1}{\log_2 N} - \dfrac{1}{\log_3 N} + \dfrac{1}{\log_4 N} - \cdots - \dfrac{1}{\log_{25} N}$, $N > 1$.

$$T_n = \log_N 2 - \log_N 3 + \log_N 4 - \log_N 5 + \cdots - \log_N 25$$
$$= \log_N 2 \cdot 4 \cdot 6 \cdot \cdots \cdot 24 - \log_N 3 \cdot 5 \cdot 7 \cdot \cdots \cdot 25$$
$$= \log_N 2 \cdot 4 \cdot 6 \cdot \cdots \cdot 24 -$$
$$\qquad\qquad \log_N \frac{(3 \cdot 5 \cdot 7 \cdot \cdots \cdot 25)(2 \cdot 4 \cdot 6 \cdot \cdots \cdot 24)}{2 \cdot 4 \cdot 6 \cdot \cdots \cdot 24}$$
$$= \log_N 2^{12} \cdot 12! - \log_N \frac{25!}{2^{12} \cdot 12!}$$
$$= \log_N \frac{(2^{12} \cdot 12!)^2}{25!}$$

19 Combinations and Probability: Choices and Chances

19-1 *Suppose that a boy remembers all but the last digit of his friend's telephone number. He decides to choose the last digit at random in an attempt to reach him. If he has only two dimes in his pocket (the price of a call is 10¢), find the probability that he dials the right number before running out of money.*

The probability of dialing correctly the first time is $\frac{1}{10}$. The probability of failing on the first attempt and succeeding on the second attempt is $\frac{9}{10} \cdot \frac{1}{9} = \frac{1}{10}$. The required probability is, therefore, $\frac{1}{10} + \frac{1}{10} = \frac{2}{10} = \frac{1}{5}$.

19-2 *In a certain town there are 10,000 bicycles, each of which is assigned a license number from 1 to 10,000. No two bicycles have the same number. Find the probability that the number on the first bicycle one encounters will not have any 8's among its digits.*

An 8 can occur in the unit's place, the ten's place, the hundred's place, and the thousand's place, each with a probability of .1, and so the probability of a digit other than 8 in any one of these positions is .9. The probability that a digit other than 8 occurs in all four positions is $(.9)^4 = .6561$, and this is the probability that there will be no 8's in the license number.

19-3 *Suppose Flash and Streak are equally strong Ping-pong players. Is it more probable that Flash will beat Streak in 3 games out of 4, or in 5 games out of 8?*

For 3 games out of 4, the probability is $\binom{4}{3}\left(\frac{1}{2}\right)^3\left(\frac{1}{2}\right) = \frac{4}{16} = \frac{8}{32}$, where $\binom{4}{3}$ means $\frac{4!}{3!1!}$ and, in general, $\binom{n}{k}$ means $\frac{n!}{k!(n-k)!}$, $k \leq n$. For 5 games out of 8, the probability is $\binom{8}{5}\left(\frac{1}{2}\right)^5\left(\frac{1}{2}\right)^3 = \frac{7}{32}$. Therefore, the more probable is 3 games out of 4.

19-4 *Show that in a group of seven people it is impossible for each person to know reciprocally only three other persons.*

The number of groupings of 3 elements from 7 elements is $\binom{7}{3} = \frac{7!}{3!4!} = 35$. Since, however, in this case the relation is reciprocal, only half as many groupings are required. But $\frac{1}{2}(35) = 17\frac{1}{2}$ groups is an obvious impossibility.

19-5 *At the conclusion of a party, a total of 28 handshakes was exchanged. Assuming that each guest was equally polite toward all the others, that is, each guest shook hands with each of the others, find the number of guests, n, at the party.*

Represent the number of guests by n. Then $\binom{n}{2} = 28$; that is, $\frac{n(n-1)}{1\cdot 2} = 28$, $n^2 - n - 56 = 0$, $(n-8)(n+7) = 0$; $n = 8$.

If we designate the guests as A, B, C, D, E, F, G, H, the 28 handshakes are AB, AC, AD, AE, AF, AG, AH; BC, BD, BE, BF, BG, BH; CD, CE, CF, CG, CH; DE, DF, DG, DH; EF, EG, EH; FG, FH; GH. Note that each letter appears 7 times in this list, corresponding to the seven handshakes made by that guest.

19-6 *A section of a city is laid out in square blocks. In one direction the streets are E1, E2, . . . , E7, and perpendicular to these are the streets N1, N2, . . . , N6. Find the number of paths, each 11 blocks long, in going from the corner of E1 and N1 to the corner of E7 and N6.*

Going from A to B (see Fig. S19-6), one can choose six $(7 - 1)$ "east" streets and five $(6 - 1)$ "north" streets. At each inter-

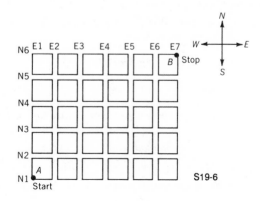

S19-6

section there is a choice of traveling along an east street or along a north street (except at the eastern boundary and the northern boundary).

Now we consider the trip composed of 11 blocks. By letting e represent an east block and n represent a north block, we can describe some possible paths as: $e, e, e, n, n, e, n, e, n, n, e$; or $e, n, e, n, e, n, e, n, e, n, e$; or $e, e, e, e, e, e, n, n, n, n, n$; or $e, e, n, n, e, e, n, n, e, e, n$; etc. . . We seek the number of different paths; that is the number of different arrangements of the e's and n's.

In all, then, there are $\binom{11}{6}$ or $\binom{11}{5}$ choices. Since $\binom{11}{6} = \frac{11!}{6!5!}$ and $\binom{11}{5} = \frac{11!}{5!6!}$, these, of course, are equal. Therefore, the number of paths is $\binom{11}{6} = 462$.

COMMENT: In general, for m east streets and n north streets, the number of paths is $\binom{m-1+n-1}{m-1} = \binom{m-1+n-1}{n-1} = \frac{(m-1+n-1)!}{(m-1)!(n-1)!}$.

19-7 *A person, starting with 64 cents, makes 6 bets, winning three times and losing three times. The wins and losses come in random order, and each wager is for half the money remaining at the time the wager is made. If the chance for a win equals the chance for a loss, find the final result.*

After each win, the amount at the time of the bet, A_i, becomes $\frac{3}{2} A_i$; and after each loss, the amount at the time of the bet, A_j, becomes $\frac{1}{2} A_j$. After three wins and three losses, in any order, the amount left is $\left(\frac{3}{2} A_{i_1}\right)\left(\frac{3}{2} A_{i_2}\right)\left(\frac{3}{2} A_{i_3}\right)\left(\frac{1}{2} A_{j_1}\right)\left(\frac{1}{2} A_{j_2}\right)\left(\frac{1}{2} A_{j_3}\right)$; that is, $\frac{27}{64} A$ where A is the original amount, since one of $A_{i_1}, A_{i_2}, A_{i_3}, A_{j_1}, A_{j_2}, A_{j_3}$ is A.

Consequently, if $A = 64$ (cents), there is a loss of $64 - 27 = 37$ (cents).

19-8 *A committee of* r *people, planning a meeting, devise a method of telephoning* s *people each and asking each of these to telephone* t *new people. The method devised is such that no person is called more than once. Find the number of people,* N, *who are aware of the meeting.*

We start with r people. The first set of telephone calls adds rs people. The second set of calls adds srt people. (See Fig. S19-8.) Therefore, $N = r + rs + rst = r(1 + s + st)$.

ILLUSTRATION: If $r = 4, s = 3, t = 2, N = 4(1 + 3 + 6) = 40$, with similar trees for r_2, r_3, r_4. The separate branches total $4(1 \times 4)$, $12(3 \times 4)$, $24(6 \times 4)$, in all, $4 + 12 + 24 = 40$.

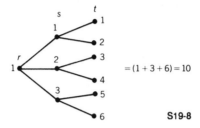

$= (1 + 3 + 6) = 10$

S19-8

19-9 *Assume there are six line segments, three forming the sides of an equilateral triangle and the other three joining the vertices of the triangle to the center of the inscribed circle. It is required that the six segments be colored so that any two with a common point must have different colors. You may use any or all of 4 colors available. Find the number of different ways to do this.*

CASE I: Three colors are used. (See Fig. S19-9.) We can choose the three colors in $\binom{4}{3} = \frac{4!}{3!1!} = 4$ (ways). Designate the colors as C_1, C_2, C_3, C_4. One possibility is $4C_1, 5C_2, 6C_3, 2C_1, 1C_2, 3C_3$. Any permutation of these subscripts will yield the same pattern.

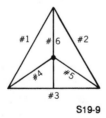

S19-9

CASE II: Four colors are used. We can choose 3 out of the 4 colors in 4 ways for segments 1, 2, 3, and the fourth color for either of the segments 4, 5, or 6. The colors of the remaining two segments are then determined. That is, there is only one way to assign the colors to the remaining two segments. Hence, when four colors are used, the six segments may be colored in $4 \cdot 3 = 12$ ways. The total number of different colorings is, therefore, $4 + 4 = 8$.

19-10 *A set of six points is such that each point is joined by either a blue string or a red string to each of the other five. Show that there exists at least one triangle completely blue or completely red.*

We may consider the six given points as the vertices of a six-sided polygon with nine diagonals since the number of diagonals of an n-sided polygon is $d = \dfrac{n(n - 3)}{2}$. Alternatively, we may reason as follows: there are 6 points to be joined by straight line segments, and this can be done in $\dbinom{6}{2} = \dfrac{6!}{2!4!} = 15$ ways; of the 15 segments, 6 are sides and 9 are diagonals. (See Fig. S19-10.)

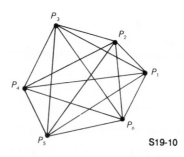

S19-10

Each triangle consists of either 2 sides and 1 diagonal, or 1 side and 2 diagonals, or 0 sides and 3 diagonals.

When there are more than 8 red or more than 8 blue there will be at least one single color triangle.

The other possibilities in tabular form are as follows.

	Red	Blue	Result
(1)	6s, 2d	0s, 7d	at least 1 blue triangle with 3 diagonals
(2)	5s, 3d	1s, 6d	at least 1 blue triangle with 3 diagonals
(3)	4s, 4d	2s, 5d	at least 1 blue triangle with 3 diagonals
(4)	3s, 5d	3s, 4d	at least 1 blue triangle with 3 diagonals
(5)	2s, 6d	4s, 3d	at least 1 blue triangle with 3 diagonals
(6)	1s, 7d	5s, 2d	at least 1 red triangle with 3 diagonals
(7)	0s, 8d	6s, 1d	at least 1 red triangle with 3 diagonals

Explanation of (1). If red strings are used for the 6 sides and 2 of the diagonals, and blue strings are used for the remaining 7 diagonals, then there is at least 1 blue triangle all of whose sides are diagonals. Similar interpretations are to be given to the other six cases.

19-11 *Each face of a cube is to be painted a different color, and six colors of paint are available. If two colorings are considered the same when one can be obtained from the other by rotating the cube, find the number of different ways the cube can be painted. [If the center of the cube is at the origin* (0, 0, 0) *the rotations are about the* x-*axis, or the* y-*axis, or the* z-*axis through multiples of* 90°.]

We can start with any one of the faces, say the top, and for it, we have a choice of 6 colors. We can then paint any one of the four side faces, say the front, and for it, there is a choice of 5 colors. This makes $6 \cdot 5 = 30$ different ways the cube can be painted.

Number the faces as follows: top #1, bottom #2, front #3, rear #5, right #4, left #6. If, instead of painting #1 and #3, we chose #1 and #4, the result would not constitute a different pattern since a 90° turn brings #4 into position #3. The same is true for faces #5 and #6. If, instead of the top, we choose the bottom, the pattern is not different since two 90° turns bring the bottom face into the top position. All other changes may be analyzed in a similar manner.

19-12 *An* 8 × 8 *checkerboard is placed with its corners at* (0, 0), (8, 0), (0, 8), *and* (8, 8). *Find the number of distinguishable non-square rectangles, with corners at points with integer coordinates, that can be counted on the checkerboard.*

It is just as easy (except for an actual count) to solve the general case of the $n \times n$ checkerboard.

The total number of rectangles is $S_1 = \left(\frac{1}{2}n(n + 1)\right)^2$ since, for each row and each column, there are rectangles 1 × 1, 1 × 2, 1 × 3, . . . , 1 × n, a total of $\frac{1}{2}n(n + 1)$ for each row and each column.

The total number of squares $S_2 = 1^2 + 2^2 + \cdots + n^2 = \frac{1}{6}n(n + 1)(2n + 1)$ (see Appendix VII), since there are n^2

squares each 1×1, $(n-1)^2$ squares each $2 \times 2, \ldots, 2^2$ squares each $(n-1) \times (n-1)$, and 1^2 square $n \times n$.

Therefore, the number of non-square rectangles is

$$S_3 = S_1 - S_2 = \left[\frac{1}{2}n(n+1)\right]^2 - \frac{1}{6}n(n+1)(2n+1)$$

$$= \frac{1}{12}(n-1)(n)(n+1)(3n+2).$$

For $n = 8$, $S_3 = \frac{1}{12} \cdot 7 \cdot 8 \cdot 9 \cdot 26 = 1092$.

19-13 *A group of* 11 *scientists are working on a secret project, the materials of which are kept in a safe. In order to permit the opening of the safe only when a majority of the group is present, the safe is provided with a number of different locks, and each scientist is given the keys to certain locks. Find the number of locks,* n_1, *required, and the number of keys,* n_2, *each scientist must have.*

The number of six-groupings of 11 items is $\binom{11}{6}$. Therefore, $\binom{11}{6} = 462$ locks are needed.

For any one scientist, his 5 companions can be chosen in $\binom{10}{5}$ ways. Therefore, each scientist needs $\binom{10}{5} = 252$ keys.

Following is an illustration with smaller numbers, say 4 scientists with 3 present at any one time. Then $\binom{4}{3} = 4$ locks are needed. For any one scientist, his companions can be chosen in $\binom{2}{1} = 2$ ways so that each scientist needs 2 keys. If we designate the scientists as S_1, S_2, S_3, and S_4, then S_1 can be provided with keys to locks 1 and 2, S_2, with keys to locks 1 and 3, S_3, with keys to locks 2 and 4, and S_4, with keys to locks 3 and 4.

All four locks can be opened with S_1, S_2, and S_3 present, with S_1, S_2, and S_4 present, with S_1, S_3, and S_4 present, and with S_2, S_3, and S_4 present.

20   An Algebraic Potpourri

20-1 *If we define* $(2n + 1)!$ *to mean the product* $(1)(2)(3) \cdots (2n + 1)$ *and* $(2n + 1)!!$ *to mean the product* $(1)(3)(5) \cdots (2n + 1)$, *express* $(2n + 1)!!$ *in terms of* $(2n + 1)!$.

$$(2n + 1)!! = (1)(3)(5) \cdots (2n + 1) = \frac{(1)(2)(3) \cdots (2n)(2n + 1)}{(2)(4)(6) \cdots (2n)}$$

Since $(2)(4)(6) \cdots (2n) = 2(1)2(2)3(2) \cdots n(2)$, it follows that $(2)(4)(6) \cdots (2n) = 2^n[(1)(2)(3) \cdots (n)] = 2^n \cdot n!$.

Therefore, $(2n + 1)!! = \dfrac{(2n + 1)!}{2^n \cdot n!}$.

20-2 *If* a, b, c *are three consecutive odd integers such that* a < b < c, *find the value of* $a^2 - 2b^2 + c^2$.

METHOD I: Since $b = \dfrac{a + c}{2}$,

$$a^2 - 2b^2 + c^2 = a^2 - 2\left(\frac{a + c}{2}\right)^2 + c^2 = \frac{(c - a)^2}{2} \cdot$$

But $c - a = 4$ (Why?), $\therefore a^2 - 2b^2 + c^2 = \dfrac{4^2}{2} = 8$.

METHOD II: $c - a = 4$ and $c + a = 2b$,

$c^2 - 2ac + a^2 = 16$, and $c^2 + 2ac + a^2 = 4b^2$.

By addition, $2(a^2 + c^2) = 4b^2 + 16$,

$a^2 + c^2 = 2b^2 + 8$, and $a^2 - 2b^2 + c^2 = 8$.

20-3 *At the endpoints* A, B *of a fixed segment of length* L, *lines are drawn meeting in* C *and making angles* α, 2α, *respectively, with the given segment. Let* D *be the foot of altitude* $\overline{CD}$ *and let* x *represent the length of* $\overline{AD}$. *Find the limiting value of* x *as* α *decreases towards zero; that is, find* $\lim\limits_{\alpha \to 0} x$.

$x \tan \alpha = (L - x) \tan 2\alpha = (L - x)\dfrac{2 \tan \alpha}{1 - \tan^2 \alpha} \therefore x = \dfrac{2L}{3 - \tan^2 \alpha}$

As $\alpha \to 0$, $\tan x \to 0$ $\therefore \lim\limits_{\alpha \to 0} x = \dfrac{2}{3} L$.

20-4 *Find the set of integers* n $\geq$ 1 *for which* $\sqrt{n - 1} + \sqrt{n + 1}$ *is rational.*

If $\sqrt{n - 1}$ is rational, let $n - 1 = k^2$; then $n + 1 = k^2 + 2$. But $k^2 + 2$ cannot be the square of an integer. (Why?) Similarly, if we assume that $n + 1 = k^2$, we find that $\sqrt{n - 1}$ is not rational.

Therefore, the required set is the empty set.

Challenge 1 *Solve the problem for* $\sqrt{n - k} + \sqrt{n + k}$ *where* k *is an integer such that* $2 \leq k \leq 8$.

ANSWER: $k = 4, n = 5; k = 6, n = 10; k = 8, n = 17$

Challenge 2 *For what positive integer values of* n *is* $\sqrt{4n - 1}$ *rational?*

ANSWER: None. (See proof below.)

PROOF: Suppose $\sqrt{4n - 1}$ is rational. We may express $\sqrt{4n - 1}$ as the fraction $\frac{p}{q}$, in lowest terms, with p and q positive integers. Then $4n - 1 = \frac{p^2}{q^2}$ and $(4n - 1)q^2 = p^2$.

If q is even then q^2 is even. (See Lemma below.) Then p^2 is even so p is even. (See Lemma below.) Since $\frac{p}{q}$ is in lowest terms this is impossible.

If q is odd then q^2 is odd. Since $4n - 1$ is odd, p^2 is odd and, p is odd. Let $p = 2a + 1$ and $q = 2b + 1$. Then $(4n - 1)(4b^2 + 4b + 1) = 4a^2 + 4a + 1$, or $4n(4b^2 + 4b + 1) - 4b^2 - 4b - 1 = 4a^2 + 4a + 1$. When the left side of this equation is divided by 4, the remainder is -1. When the right side is divided by 4, the remainder is $+1$. Since this is a contradiction, the assumption that $\sqrt{4n - 1} = \frac{p}{q}$ is untenable. Therefore, $\sqrt{4n - 1}$ is irrational for all positive integer values of n.

LEMMA: If q is even, that is $q = 2r$, then q^2 is even. Since $q = 2r$, $q^2 = 4r^2 = 2(2r^2)$, an even number.

If q^2 is even, that is $q^2 = 2s$, then q is even. Suppose q is odd, that is $q = 2t + 1$. Then $q^2 = 4t^2 + 4t + 1 = 2(2t^2 + 2t) + 1$. Therefore, $2s = 2(2t^2 + 2t) + 1$. The left side of this equation is even while the right side is odd. This is an impossibility. The assumption that q is odd is untenable. Therefore, q is even.

20-5 *The angles of a triangle* ABC *are such that* $\sin$ B $+ \sin$ C $= 2 \sin$ A. *Find the value of* $\tan \frac{B}{2} \tan \frac{C}{2}$.

$$\sin B + \sin C = 2 \sin \frac{B+C}{2} \cos \frac{B-C}{2}$$

$$2 \sin A = 2 \sin [180° - (B + C)]$$

$$= 2 \sin (B + C) = 4 \sin \frac{B+C}{2} \cos \frac{B+C}{2}$$

$$\cos \frac{B-C}{2} = 2 \cos \frac{B+C}{2}$$

$$\cos \frac{B}{2} \cos \frac{C}{2} + \sin \frac{B}{2} \sin \frac{C}{2} = 2 \cos \frac{B}{2} \cos \frac{C}{2} - 2 \sin \frac{B}{2} \cos \frac{C}{2}$$

$$3 \sin \frac{B}{2} \sin \frac{C}{2} = \cos \frac{B}{2} \cos \frac{C}{2}; \ 3 \tan \frac{B}{2} \tan \frac{C}{2} = 1$$

$$\therefore \tan \frac{B}{2} \tan \frac{C}{2} = \frac{1}{3}$$

20-6 *Decompose* $F = \dfrac{7x^3 - x^2 - x - 1}{x^3(x - 1)}$ *into the sum of fractions with constant numerators.*

Divisors of the denominator $x^3(x - 1)$ are x^3, x^2, x, and $x - 1$.

Let $F = \dfrac{7x^3 - x^2 - x - 1}{x^3(x - 1)} = \dfrac{A}{x^3} + \dfrac{B}{x^2} + \dfrac{C}{x} + \dfrac{D}{x - 1}$ be an identity in x.

Then $7x^3 - x^2 - x - 1 = A(x - 1) + Bx(x - 1) + Cx^2(x - 1) + Dx^3$ holds for all x.

Let $x = 0$; then $-1 = -A$, $A = 1$.

Let $x = 1$; then $D = 4$.

Let $x = -1$; then $-8 = -2 + 2B - 2C - 4$, and $B - C = -1$.

Let $x = 2$; then $49 = 1 + 2B + 4C + 32$, and $B + 2C = 8$.

Thus, $B = 2$ and $C = 3$. $F = \dfrac{1}{x^3} + \dfrac{2}{x^2} + \dfrac{3}{x} + \dfrac{4}{x - 1}$

20-7 *On a transcontinental airliner there are 9 boys, 5 American children, 9 men, 7 foreign boys, 14 Americans, 6 American males, and 7 foreign females. Find the number of people on the airliner.*

The table shown is constructed as follows:

(1) Enter 7   at the intersection of row B and Column F.
(2) Enter 2   at the intersection of row B and Column A.
(3) Enter 3   at the intersection of row G and Column A.
(4) Enter 4   at the intersection of row M and Column A.
(5) Enter 5   at the intersection of row W and Column A.

(6) Enter 5   at the intersection of row *M* and Column *F*.
(7) Enter *X*   at the intersection of row *G* and Column *F*.
(8) Enter $7 - X$ at the intersection of row *W* and Column *F*.

		A (American)	*F* (Foreign)
Children	*B* (boys)	2	7
	G (girls)	3	*X*
Adults	*M* (men)	4	5
	W (women)	5	$7 - X$
		14	19

The total is $14 + 19 = 33$.

NOTE: The number of foreign girls is not determined except to say that it can vary from zero to seven.

20-8 *If* a *and* b *are positive integers and* b *is not the square of an integer, find the relation between* a *and* b *so that the sum of* $a + \sqrt{b}$ *and its reciprocal is integral.*

Let $a + \sqrt{b} + \dfrac{1}{a + \sqrt{b}} = m$; then $ma + m\sqrt{b} = (a^2 + b + 1) + 2a\sqrt{b}$.

$\therefore m\sqrt{b} = 2a\sqrt{b}$ so that $m = 2a$, and $ma = a^2 + b + 1$.

$2a^2 = ma = a^2 + b + 1 \therefore a^2 = b + 1$

Challenge *Solve the problem so that the sum is rational but not integral.*

Let $S = a + \sqrt{b} + \dfrac{1}{a + \sqrt{b}} = a + \sqrt{b} + \dfrac{1}{a + \sqrt{b}} \cdot \dfrac{a - \sqrt{b}}{a - \sqrt{b}}$

$= \dfrac{a^3 - ab + a + a^2\sqrt{b} - b\sqrt{b} - \sqrt{b}}{a^2 - b}$.

For S to be rational the numerator must be an integer, so that $a^2 - b - 1 = 0$; that is, $a^2 = b + 1$. But we have already established that when $a^2 = b + 1$, S is an integer. Therefore, the Challenge has no solution.

20-9 *Find the simplest form for* $R = \sqrt{1 + \sqrt{-3}} + \sqrt{1 - \sqrt{-3}}$.

Let $\sqrt{a} + \sqrt{-b} = \sqrt{1 + \sqrt{-3}}$, with $a > 0$, $b > 0$. Then $a - b + 2i\sqrt{ab} = 1 + i\sqrt{3}$ where $i = \sqrt{-1}$. So $a - b = 1$ and $2\sqrt{ab} = \sqrt{3}$. Squaring each of the last two equations, we have $a^2 - 2ab + b^2 = 1$ (I) and $4ab = 3$ (II).

By addition of (I) and (II) we have $a^2 + 2ab + b^2 = 4$, so that $a + b = 2$. (Why is the value -2 rejected?)

We now solve the pair of equations $a - b = 1$ and $a + b = 2$ to get $a = \frac{3}{2}$ and $b = \frac{1}{2}$, and, hence, $\sqrt{1 + \sqrt{-3}} = \sqrt{\frac{3}{2}} + \sqrt{-\frac{1}{2}}$.

In a similar manner we obtain $\sqrt{1 - \sqrt{-3}} = \sqrt{\frac{3}{2}} - \sqrt{-\frac{1}{2}}$.

Therefore, $R = \sqrt{1 + \sqrt{-3}} + \sqrt{1 - \sqrt{-3}} = \sqrt{\frac{3}{2}} + \sqrt{-\frac{1}{2}} + \sqrt{\frac{3}{2}} - \sqrt{-\frac{1}{2}} = 2\sqrt{\frac{3}{2}} = \sqrt{6}$.

20-10 *Observe that the set* $\{1, 2, 3, 4\}$ *can be partitioned into subsets* T_1 $\{4, 1\}$ *and* T_2 $\{3, 2\}$ *so that the subsets have no element in common, and the sum of the elements in* T_1 *equals the sum of the elements in* T_2. *This cannot be done for the set* $\{1, 2, 3, 4, 5\}$ *or the set* $\{1, 2, 3, 4, 5, 6\}$. *For what values of* n *can a subset of the natural numbers* $S_n = \{1, 2, 3, \ldots, n\}$ *be so partitioned?*

The sum of the elements of S_n is $T_n = \frac{1}{2}n(n + 1)$ (see Appendix VII). In order that $\frac{1}{2}T_n$ be integral, the number $\frac{1}{2}\left[\frac{1}{2}n(n + 1)\right] = \frac{1}{4}n(n + 1)$ must be integral. Consequently, either n is a multiple of 4 or $n + 1$ is a multiple of 4; that is, either $n = 4k$ or $n + 1 = 4k$, where $k = 1, 2, 3, \ldots$.

ILLUSTRATION: When $k = 1$, $n = 4$ or 3. The case $n = 4$ is the favorable case given in the problem. The case $n = 3$ is somewhat trivial, namely, $\{1, 2, 3\}$ with $T_1 = \{1, 2\}$ and $T_2 = \{3\}$.

The two unfavorable illustrations given in the problem are, respectively, cases of $n = 4k + 1$ and $n = 4k + 2$ with $k = 1$.

20-11 *Suppose it is known that the weight of a medallion,* X *ounces, is represented by one of the integers* 1, 2, 3, $\ldots$, N. *You have available a balance and two different weights, each with an integral number of ounces, represented by* W_1 *and* W_2. *Let* S $= N + W_1 + W_2$. *Find the value of* S *for the largest possible value of* N *that can be determined with the given conditions.*
(For this problem we are indebted to Professor M. I. Aissen, Fordham University.)

Suppose $W_1 = 2$. If $X < 2$, then $X = 1$. If $X = 2$, then $X = 2$.
If $X > 2$, go to the next step.

Suppose $W_2 = 6$. If $X + 2 < 6$, then $X = 3$. If $X + 2 = 6$,
then $X = 4$.
If $X + 2 > 6$, go to the next step.
If $X < 6$, then $X = 5$. If $X = 6$, $X = 6$.
If $X > 6$, go to the next step.
If $X < 6 + 2$, then $X = 7$. If $X = 6 + 2$, then
$X = 8$.
If $X > 6 + 2$, then $X = 9$, the largest deter-
minable value. Therefore, $N + W_1 + W_2 =
9 + 2 + 6 = 17$.

20-12 *If* M *is the midpoint of line segment* $\overline{AB}$, *and point* P *is between* M *and* B, *and point* Q *is beyond* B *such that* $QP^2 = QA \cdot QB$, *show that, with the proper choice of units, the length of* $\overline{MP}$ *equals the smaller root of* $x^2 - 10x + 4 = 0$. (A, M, B, P, *and* Q *are collinear.*)

Since $x^2 - 10x + 4 = 0$,
$x^2 - 10x + 25 = 21. \therefore (5 - x)^2 = 7 \cdot 3, x = 5 - \sqrt{21}$
If we let $QB = 3$ and $QA = 7$ (Fig. S20-12),
then $BA = 4$, $BM = MA = 2$. Represent MP by x.
Then $QP = (QB + BM) - MP = (3 + 2) - x$.

S20-12

The geometric relation $QP^2 = QA \cdot QB$ corresponds to the
algebraic relation $(5 - x)^2 = 7 \cdot 3$
$QP = QM - MP$ corresponds to $5 - x$ so that $MP = x =
5 - \sqrt{21}$.

20-13 *Consider the lattice where* R_i *is the i-th row and* C_j *is the j-th column, i, j, = 1, 2, 3, ..., in which all the entries are natural numbers. Find the row and column for the entry 1036.*

The initial entry in R_{n+1} is $\frac{1}{2}n(n + 1) + 1$, $n = 0, 1, 2, 3, \ldots$,
and the number of entries is $n + 1$; R_1 contains the single entry 1.
(See Fig. S20-13.) We therefore set $1036 = \frac{1}{2}n(n + 1) + 1$ from

which we get $n^2 + n - 2070 = 0$. Solving by factoring, we have $(n + 46)(n - 45) = 0$ so that $n = 45$. Hence, the initial entry R_{46} is 1036 so that 1036 is at the intersection of row 46 and column 1.

In the case of an entry such as 1036, we successfully, and fortunately, found a positive integral root for the equation involving n. Is this the case with an entry such as 212?

	C_1	C_2	C_3	•	•	•	
R_1	1						
R_2	2	3					
R_3	4	5	6				
•	•	•	•	•			
•	•	•	•	•	•		
•	•	•	•	•	•	•	

S20-13

Following the procedure used before, we set $212 = \frac{1}{2} n(n + 1) + 1$. From this we get $n^2 + n - 422 = 0$, and this equation does not have a positive integral root. However, we do find that $20 < n < 21$. Since the initial entry for R_{21} is $\frac{1}{2}(20)(21) + 1 = 211$ and the initial entry for R_{22} is $\frac{1}{2}(21)(22) + 1 = 232$, we see that 212 is at the intersection of row 21 and column 3.

20-14 *Express* $P(c) = c^6 + 10c^4 + 25c^2$ *as a polynomial of least positive degree when c is a root of* $x^3 + 3x^2 + 4 = 0$.

METHOD I: Since c is a root of $x^3 + 3x^2 + 4 = 0$, we have $c^3 + 3c^2 + 4 = 0$ (I). We multiply (I) successively by c, c^2, and c^3 to obtain $c^4 + 3c^3 + 4c = 0$ (II), $c^5 + 3c^4 + 4c^2 = 0$ (III), and $c^6 + 3c^5 + 4c^3 = 0$ (IV). From (I) we have $c^3 = -3c^2 - 4$ (V), and from (II) we have $c^4 = -3c^3 - 4c$ (VI). Substituting (V) into (VI), we get $c^4 = 9c^2 - 4c + 12$ (VII). From (III) we get $c^5 = -3c^4 - 4c^2$ (VIII) so that $c^5 = -31c^2 + 12c - 36$ [after substituting (VII) into (VIII)]. Finally, by similar operations, we get $c^6 = -3c^5 - 4c^3 = 105c^2 - 36c + 124$ (IX).

The sum of (IX) and 10 times (VII) and $25c^2$ is, therefore, $c^6 + 10c^4 + 25c^2 = 220c^2 - 76c + 244$.

METHOD II: $P(c) = Q(c)(c^3 + 3c^2 + 4) + R(c)$ where $Q(c)$ is the quotient obtained when $P(c)$ is divided by $c^3 + 3c^2 + 4$ and $R(c)$, of degree less than 3, is the remainder. But $c^3 + 3c^2 + 4 = 0$. Therefore, $P(c) = R(c)$, which is found by actual division to be $220c^2 - 76c + 244$.

20-15 *Let* $S = \dfrac{b_0 + b_1 x + \cdots + b_n x^n}{1 - x} = r_0 + r_1 x + \cdots + r_n x^n$ *be an identity in* x. *Express* r_n *in terms of the given* b's.

Multiply both sides of the identity by $1 - x$.
$$b_0 + b_1 x + \cdots + b_n x^n = r_0 + r_1 x + r_2 x^2 + \cdots + r_n x^n$$
$$- r_0 x - r_1 x^2 - \cdots - r_{n-1} x^n - r_n x^{n+1}$$
$\therefore r_0 = b_0;\ r_1 - r_0 = b_1$ so that $r_1 = b_0 + b_1$, and so forth.
$$r_n = b_0 + b_1 + \cdots + b_n$$

20-16 *Find the numerical value of the infinite product* P *whose factors are of the form* $\dfrac{n^3 - 1}{n^3 + 1}$, *where* n = 2, 3, 4, ...

$n^3 - 1 = (n - 1)(n^2 + n + 1)$ and
$n^3 + 1 = (n + 1)(n^2 - n + 1)$ (See Appendix VII.)
$$P = \frac{1 \cdot 7}{3 \cdot 3} \cdot \frac{2 \cdot 13}{4 \cdot 7} \cdot \frac{3 \cdot 21}{5 \cdot 13} \cdots \frac{(n - 1)(n^2 + n + 1)}{(n + 1)(n^2 - n + 1)}$$
$$\cdot \frac{n(n^2 + 3n + 3)}{(n + 2)(n^2 + n + 1)} \cdot \frac{(n + 1)(n^2 + 5n + 7)}{(n + 3)(n^2 + 3n + 3)} \cdots$$

Every factor in the numerator is matched with an equal factor in the denominator, with the exception of 1 and 2 in the numerator and 3 in the denominator.

Therefore, $P = \dfrac{2}{3}$.

20-17 *Express* $F = \dfrac{\sqrt[3]{2}}{1 + 5\sqrt[3]{2} + 7\sqrt[3]{4}}$ *with a rational denominator.*

METHOD I: Let $r = 1 + 5\sqrt[3]{2} + 7\sqrt[3]{4}$.
$$(r - 1)^3 = (5\sqrt[3]{2} + 7\sqrt[3]{4})^3$$
$$r^3 - 3r^2 + 3r - 1 = 250 + 1050\sqrt[3]{2} + 1470\sqrt[3]{4} + 1372$$
$$= 1622 + 210(5\sqrt[3]{2} + 7\sqrt[3]{4})$$
$$= 1622 + 210(r - 1)$$
$$= 1412 + 210r$$

$$r^3 - 3r^2 - 207r = 1413, \quad r^2 - 3r - 207 = \frac{1413}{r}$$

$$\frac{1}{r} = \frac{r^2 - 3r - 207}{1413} = \frac{r(r-3) - 207}{1413}$$

$$= \frac{(1 + 5\sqrt[3]{2} + 7\sqrt[3]{4})(-2 + 5\sqrt[3]{2} + 7\sqrt[3]{4}) - 207}{1413}$$

$$= \frac{-23 + 31\sqrt[3]{2} + 6\sqrt[3]{4}}{471}$$

Therefore, $F = \dfrac{\sqrt[3]{2}}{r} = \dfrac{-23\sqrt[3]{2} + 31\sqrt[3]{4} + 12}{471}$.

METHOD II: In this method we use the identities
$m^3 - n^3 = (m - n)(m^2 + mn + n^2)$ and
$m^3 + n^3 = (m + n)(m^2 - mn + n^2)$. (See Appendix VII.)

Start with trinomial $1 + ax + bx^2$, with $x = \sqrt[3]{2}$. Then $(1 + ax + bx^2)(1 - ax) = (1 - abx^3) - (a^2 - b)x^2 = c - dx^2$ where $c = 1 - abx^3$ and $d = a^2 - b$ are both rational.

Since $c - dx^2$ is still irrational, we apply the method a second time. $(c - dx^2)(c^2 + cdx^2 + d^2x^4) = c^3 - d^3x^6$, a rational expression.

In the given problem, $a = 5$, $b = 7$, $c = -69$, $d = 18$, $c^2 = (-69)^2$, $cdx^2 = (-69)(18)\sqrt[3]{4}$, $d^2x^4 = (18)^2(4^2)$. F is rationalized by multiplying the numerator and the denominator by $(1 - 5\sqrt[3]{2})(69^2 - 69 \cdot 18\sqrt[3]{4} + 18^2 \cdot 4^2)$ or, more simply, by $(1 - 5\sqrt[3]{2})(23^2 - 23 \cdot 6\sqrt[3]{4} + 6^2 \cdot 4^2)$, where the second factor was divided by $3^2 = 9$.

With these operations performed correctly, we obtain the value of F shown in Method I.

Challenge *Express* $F = \dfrac{\sqrt[3]{2}}{1 + \sqrt[3]{2} + \sqrt[3]{4}}$ *with a rational denominator.*

This can be done by Method I or by Method II, but also by a specialized method resulting from the fact that $1 + \sqrt[3]{2} + \sqrt[3]{4}$ is a geometric series.

Since $1 + \sqrt[3]{2} + \sqrt[3]{4} = \dfrac{(\sqrt[3]{2})^3 - 1}{\sqrt[3]{2} - 1} = \dfrac{1}{\sqrt[3]{2} - 1}$,

$F = \dfrac{\sqrt[3]{2}}{\dfrac{1}{\sqrt[3]{2} - 1}} = \sqrt[3]{4} - \sqrt[3]{2}$.   (See Appendix VII.)

20-18 *Starting with the line segment from 0 to 1 (including both endpoints), remove the open middle third; that is, points $\frac{1}{3}$ and $\frac{2}{3}$ of the middle third remain. Next remove the open middle thirds of*

the two remaining segments $\left(points\ \frac{1}{9},\ \frac{2}{9},\ \frac{7}{9},\ \frac{8}{9}\ remain\ along\ with\right.$ $\left.\frac{1}{3}\ and\ \frac{2}{3}\right)$. *Then remove the open middle thirds of the four segments remaining, and so on endlessly. Show that one of the remaining points is* $\frac{1}{4}$.

In succession, starting with the point $\frac{1}{3}$, we have intervals represented by the sums s_n of the series $\frac{1}{3} - \frac{1}{9} + \frac{1}{27} - \frac{1}{81} + \cdots \pm \frac{1}{3^n}$; that is, $s_1 = \frac{1}{3}$, $s_2 = \frac{1}{3} - \frac{1}{9}$, and so forth.

$$\therefore \lim_{n \to \infty} s_n = \frac{\dfrac{1}{3}}{1 - \left(-\dfrac{1}{3}\right)} = \frac{1}{4}$$

Expressed in the base 3, the number is $\frac{1}{3} - \frac{1}{3^2} = \frac{0}{3} + \frac{2}{3^2}$ plus $\frac{1}{3^3} - \frac{1}{3^4} = \frac{0}{3^3} + \frac{2}{3^4}$ plus $\dots$, or $.020202\dots$ (base 3) or $.\overline{02}$ (base 3).

COMMENT: The set of points formed from the closed interval [0, 1] by removing first the middle third of the interval, then the middle third of each remaining interval, and so on indefinitely, is known as the Cantor Set, and also the Cantor Discontinuum. It has unusual and interesting properties.

20-19 *Write a formula that can be used to calculate the n-th digit a_n of* $N = .01001000100001\dots$, *where all the digits are either 0 or 1, and where each succeeding block has one more zero than the previous block.*

The digit "1" appears in positions 2, 5, 9, 14, $\dots$. To find a formula for generating this sequence, we may proceed as follows. The first differences of successive terms in the sequence are 3, 4, 5, $\dots$, that is, $5 - 2 = 3$, $9 - 5 = 4$, $14 - 9 = 5$, $\dots$ and the second differences are 1, 1, 1, $\dots$.

Since the second differences are constant, we try the formula $Ak^2 + Bk + C$, $k = 1, 2, 3, \dots$.

When $k = 1$ we have $2 = A + B + C$.

When $k = 2$ we have $5 = 4A + 2B + C$.

When $k = 3$ we have $9 = 9A + 3B + C$.

The common solution of this set of equations is $A = \frac{1}{2}$, $B = \frac{3}{2}$, $C = 0$. Therefore, the generating function is $\frac{1}{2}k^2 + \frac{3}{2}k$, $k = 1, 2, 3, \ldots$.

Therefore, $a_n = \begin{cases} 1 & \text{when } n = \frac{1}{2}k^2 + \frac{3}{2}k, \ k = 1, 2, 3, \ldots; \\ 0 & \text{otherwise.} \end{cases}$

Challenge *Find* a_n *the* n-*th digit, of* $M = .101001000100001\ldots.$

METHOD I: Use the method shown above.

METHOD II: The digit "1" appears in positions 1, 3, 6, 10, ..., but these are the triangular numbers T_1, T_2, T_3, T_4, ... (See Appendix VII.)

Therefore, $a_n = \begin{cases} 1 & \text{when } n = \frac{1}{2}k^2 + \frac{1}{2}k, \ k = 1, 2, 3, \ldots; \\ 0 & \text{otherwise.} \end{cases}$

ANSWERS

1-1 3   *Challenge 1*: 4   *Challenge 2*: 7

1-2 See solution.

1-3 See solution.   *Challenge 1*: At least one box with 3 or more letters   *Challenge 2*: At least one box with 3 or more letters   *Challenge 3*: At least one box with 4 or more letters   *Challenge 4*: 26

1-4 See solution.   *Challenge 1*: See solution.   *Challenge 2*: 4   *Challenge 3*: 4

1-5 192 m.p.h.

1-6 $n = 9$ *Challenge*: $n = 7$

1-7 1:06 P.M.

1-8 $14\frac{3}{4}$ hours   *Challenge 1*: $9\frac{3}{4}$ hours   *Challenge 2*: 1:06 A.M.

1-9 3

1-10 54   *Challenge 1*: 70

1-11 120   *Challenge*: 248 oz.

1-12 Plan I yields \$200 more.   *Challenge*: For the 5-year interval, earnings are the same.

1-13 See solution.

1-14 **(a)** $L = j + n - 1$ **(b)** $M = j + \frac{n-1}{2}$ for odd n; $M = j + \frac{n}{2} - 1$, or $j + \frac{n}{2}$ for even n *Challenge 1*: **(a)** $S = j - n + 1$ **(b)** $M = j - \frac{n-1}{2}$ for odd n; $M = j - \frac{n}{2} + 1$, or $j - \frac{n}{2}$ for even n *Challenge 2*: **(a)** $L = j + 2(n - 1)$ **(b)** $M = j + n - 1$ for odd n; $M = j + n - 2$, or $j + n$ for even n *Challenge 3*: Same answers as for Challenge 2   *Challenge 4*: $j + \frac{3n}{4} - 1$; $j - \frac{n}{4}$; $j + \frac{3n}{2} - 2$; $j + \frac{3n}{2} - 2$

1-15 $|x - y| = \max(x, y) - \min(x, y)$ *Challenge 1*: Yes   *Challenge 2*: $\min(x, y) = \dfrac{x + y}{2} - \dfrac{|x - y|}{2}$

1-16 (a) $x = x^+ - x^-$ (b) $|x| = x^+ + x^-$ (c) $x^+ = \frac{1}{2}(|x| + x)$
(d) $x^- = \frac{1}{2}(|x| - x)$

1-17 1 or 0   *Challenge 1*: (a) $2y - 1$ or $2[y]$ (b) $-2[y]$ or $1 - 2y$
Challenge 2: (a) $F = 1$ (b) $0 < F < 1$ (c) $1 < F < \infty$ *Challenge 3*: (a) $D = 0$ (b) $D = 0$ or 1 or 2 (c) $D = 0$ or 1 or 2 or 3
or 4   *Challenge 4*: $x = 3\frac{2}{3}$ *Challenge 5*: $(x + y) = (x) +$
(y) when $(x) + (y) < 1$, $(x + y) = (x) + (y) - 1$ when $(x) +$
$(y) \geq 1$

1-18 $4{:}21\frac{9}{11}$ *Challenge   1*: $7{:}54\frac{6}{11}$ *Challenge   2*: $5{:}17\frac{29}{143}$
Challenge 3: $8\frac{8}{143}$ hours

2-1 44.25   *Challenge 1*: $-17\frac{7}{12}$ *Challenge 2*: 42.5   *Challenge 3*: (a) 4.2 or (b) 4.3   *Challenge 4*: (a) $\dfrac{n + 3}{n + 2}$ (b) $\dfrac{n + 2}{n + 1}$
Challenge 5: 1   *Challenge 7*: $70\frac{7}{8}$

2-2 $4|\text{A.M.}|$ *Challenge*: No change.

2-3 No   *Challenge*: No

2-4 $n = 40$ *Challenge*: Set $m = -n$ in order to solve.

2-5 $1, \dfrac{d}{c}, \dfrac{b + d}{a + c}, \dfrac{b}{a}, \dfrac{bd}{ac}$

2-6 (a) 6 (b) 4 (c) 3 (d) 4 (e) 2   *Challenge*: (a) 5 (b) 8 (See solution.)

2-7 24   *Challenge*: 4

2-8 18   *Challenge 1*: 9   *Challenge 2*: 14

2-9 8 possibilities (See solution.)   *Challenge*: 8 possibilities (See solution.)

2-10 $k = 8$ *Challenge 1*: 25   *Challenge 2*: 7   *Challenge 3*:
60 days

2-11 0 or 4   *Challenge*: Only 1

2-12 Even integers (See solution.)

2-13 $m_1 = 2, m_2 = 3, m_3 = 1$ *Challenge 1*: See solution.

2-14 -7 *Challenge 1*: 8   *Challenge 2*: -23

2-15 1 or -1 *Challenge 1:* -7 *Challenge 2:* $m^6 + 1$, $m^6 + 1$, $m^3 + 1$

2-16 $N = 51{,}051$, or $69{,}069$

2-17 No

2-18 24; 93,324

2-19 $k = 1, 2, 19$ *Challenge:* $k = 3, 8$

2-20 r when r is odd, $2r$ when r is even   *Challenge 1:* $2r$ when r is odd, r when r is even   *Challenge 2:* r when r is odd

2-21 $\sqrt{n}$

3-1 $y_6 = x$, $y_{100} = x$, $y_{501} = \dfrac{x+1}{x-1}$ *Challenge 1:* 2   *Challenge 2:* 3   *Challenge 3:* Undefined   *Challenge 4:* Undefined.

3-2 (a) $s_1 s_2 - 2s_1 - 2s_2 + 4$ (b) $2s_1 + 2s_2 - 4$, or $2[(s_1 - 1) + (s_2 - 1)]$ (c) $s_1 s_2$ *Challenge:* $N(I) = s^2 - 4s + 4$, $N(B) = 4s - 4$, $N = s^2$, where s is the number of lattice points in the side of the square.

3-3 The change in the value of p is 13.5 mm. (decrease).   *Challenge:* 715 mm., 760 mm.

3-4 $1.65   *Challenge 1:* $1.85   *Challenge 2:* $2.05   *Challenge 3:* See solution.

3-5 $x = -\dfrac{1}{2}$, $y = -1$ *Challenge 1:* $x = \dfrac{3\sqrt{2}}{3 - \sqrt{2}}$, $y = \sqrt{2}$, or $x = \dfrac{-3\sqrt{2}}{3 + \sqrt{2}}$, $y = -\sqrt{2}$ *Challenge 2:* $x = 2$, $y = 1$, and $x = -\dfrac{2}{3}$, $y = -1$

3-6 $n = 12$ *Challenge:* $n = 6k - 1$, $k = 1, 2, 3, \ldots$.

3-7 $33\frac{1}{3}$ *Challenge:* $100 - 100\left(1 + \dfrac{a+b}{100} + \dfrac{ab}{100^2}\right)^{-1}$

3-8 $x = \dfrac{b}{p - m}$ *Challenge 1:* $p > m$ *Challenge 2:* Fixed overhead (expense)   *Challenge 3:* $x = \dfrac{100 + b}{p - m}$ *Challenge 4:* $x = \dfrac{D + b}{p - m}$ *Challenge 5:* b dollars

3-9 60% *Challenge 1:* 50% *Challenge 2:* $66\frac{2}{3}\%$

3-10 $n_1 : n_2 = 1 : 2$ *Challenge:* See solution.

3-11 See solution.

3-12 $x = \dfrac{1}{2}(a + b)$ *Challenge 1:* $x = 1\frac{1}{2}$ *Challenge 2:* $x = \dfrac{1}{2}$ *Challenge 3:* $x = -1$ or 1 *Challenge 4:* $x = -\dfrac{1}{2}$ or $\dfrac{3}{4}$ *Chal-*

lenge 5: $x = \frac{3}{4}$ *Challenge 6*: $x = 2$ or $\frac{4}{5}$ *Challenge 7*:
$x = 0$ or 2

3-13 Smith and Jones cannot get together.   *Challenge 1*: 65 days for first reunion

3-14 No   *Challenge 1*: **(a)** No **(b)** Likely, but there is no certainty. *Challenge 2*: One possibility is 1 quarter, 2 nickels; a second possibility is 1 dime, 2 nickels.

3-15 12 ways   *Challenge*: Yes

3-16 3 ways

4-1 See solution.

4-2 0   *Challenge 1*: 1   *Challenge 2*: 0, 1, 2   *Challenge 3*: 1, 2, 3

4-3 See solution.   *Challenge 1*: 11   *Challenge 2*: 19

4-4 All bases $b > 4$ *Challenge 1*: 111 (base 4)   *Challenge 2*: 12

4-5 See solution.

4-6 9   *Challenge 1*: Yes, Yes   *Challenge 2*: 495

4-7 110.11 (base 2)   *Challenge 1*: 10011.10101 (base 2)   *Challenge 2*: No (See solution.)

4-8 $r = 9$ *Challenge*: $r = 6$ or 10

4-9 $B(.1, 0)$, $D(.01, 0)$, $E(.02, 0)$ *Challenge*: $C(.2, 0)$, $F(.21, 0)$, $G(.22, 0)$

4-10 See solution.   *Challenge 1*: Eighth   *Challenge 3*: See solution.

5-1 The empty set   *Challenge*: The empty set

5-2 $x = 3$, y, any real value except $3\frac{1}{2}$; or x, any real value, $y = 4$ *Challenge*: $y = 1$, x, any real number, or $x = 3$, y any real value except $3\frac{1}{2}$ or $\frac{2}{7}$

5-3 $3 - 2\sqrt{2} < x < 3 + 2\sqrt{2}$ *Challenge*: $b - a = 6$

5-4 $x = 1$ *Challenge*: $x = \frac{1}{3}$

5-5 $a > 3, b = 2$ *Challenge*: $a \leq 3, b = 2$

5-6 $N = 35k + 16$, $k = 0, 1, 2, 3, \ldots$ *Challenge 1*: $N = 35m + 22$, $m = 0, 1, 2, 3, \ldots$ *Challenge 2*: See solution.

5-7 20   *Challenge 1*: $p = \frac{6b}{5}$, $s = a + \frac{3b}{5}$ *Challenge 2*: $p = \frac{r_1 r_2}{r_1 r_2 - 1}(b)$, $s = a + \frac{r_2}{r_1 r_2 - 1}(b)$

5-8 $\dfrac{250}{\sqrt{5}}$ seconds

5-9 $x = 12, y = 3$ *Challenge*: Yes

5-10 $\left\{-\dfrac{11}{2}\right\}$ *Challenge 1*: $\left\{-\dfrac{13}{2}\right\}$ *Challenge 2*: $\left\{-a - \dfrac{3}{2}\right\}$

5-11 Least possible, 17; largest possible, 19   *Challenge 1*: Smallest combination, 6-6-6   *Challenge 2*: Smallest combination, 6-6-6   *Challenge 3*: See solution.   *Challenge 4*: See solution.

5-12 Questionable answer: 3 item A, 2 item B, 15 item C (See solution.)

5-13 8, 16, 3, 48   *Challenge 1*: 6, 12, 3, 27   *Challenge 2*: 6, 10, 4, 16, 64

5-14 $6\dfrac{2}{3}$ miles

5-15 200 miles

5-16 25 f.p.s.

5-17 $x = 12$

5-18 113, 223   *Challenge*: 17, 31, 53

5-19 5:20 P.M.   *Challenge 1*: 4:48 P.M.   *Challenge 2*: Change $\dfrac{1}{8}$ to $\dfrac{11}{76}$ *Challenge 3*: The professor is right.

5-20 $z = kw$, $y = z + w$, $x = z + \dfrac{z^2}{w}$, w an integer, $k = \pm 1, \pm 2,$ ... (See solution.)

5-21 See solution.

6-1 21   *Challenge 1*: 36   *Challenge 2*: See solution   *Challenge 3*: $f(10) = 55$, $f(100) = 5050$ *Challenge 4*: $n = 9$ *Challenge 5*: $n = 2$

6-2 (a) $f(n) = 1 - n$ (b) $f(n) = n + 1$ (c) $f(n) = 1 - 2n$ (d) $f(n) = n + \dfrac{3}{4}$ (e) $f(n) = n - \dfrac{3}{4}$ *Challenge*: $f(n) = 2n - 3$

6-3 5, 12, 13   *Challenge*: 5, 12, 13

6-4 4, 3; 5, 3; 5, 4

6-5 $\left[0, \dfrac{1}{2}\right]$ *Challenge*: See solution.

6-6 20

6-7 $4n$

6-8 $\dfrac{1}{2}(n^2 + n + 2)$ *Challenge 1*: $\dfrac{1}{2}(r^2 + r + 2) + k(r + 1)$ *Challenge 2*: $\dfrac{1}{2}n(n + 1)$ *Challenge 3*: $(k_1 + 1)(k_2 + 1)$

Challenge 4: $k_1 + k_2 + 1$, $k_1k_2 + 2(k_1 + k_2 + 1)$ *Challenge 5*: $\frac{1}{2}(n^2 + n - 2)$

6-9 32   *Challenge*: 24

7-1 $k = 0$
7-2 $X^2 + Y^2 = 2$
7-3 $\frac{21}{37}$
7-4 $\frac{3k^2}{16}$ *Challenge 1*: $\frac{k^2}{12}$ *Challenge 2*: See solution.
7-5 See solution.
7-6 190
7-7 See solution.   *Challenge*: See solution.
7-8 $x = 11, y = -2$
7-9 $n = 2$ and $n = 3$
7-10 $x = 4, y = 5$
7-11 $x = 1.1$, $y = 0.9$, $z = -0.5$; $x = \frac{32}{30}$, $y = \frac{25}{30}$, $z = -\frac{12}{30}$
7-12 $x = \frac{11}{10}$, $y = \frac{5}{6}$ (See solution.)
7-13 maximum 17, minimum $-\frac{1}{2}$
7-14 10 item A, 30 item B, 60 item C

8-1 $x = -\frac{8}{9}$
8-2 (a) 2, -2 (b) 2   *Challenge*: (a) none (b) -2
8-3 (a) 18 (b) One possibility is P_1 (0, 5), P_2 (2, 7); a second possibility is $P_1 \left(\frac{1}{2}, \frac{11}{2}\right)$, $P_2 \left(\frac{3}{2}, \frac{13}{2}\right)$.
8-4 7   *Challenge 1*: 14   *Challenge 2*: 3   *Challenge 3*: 6
8-5 3   *Challenge*: 3
8-6 All integral values of x and p *Challenge*: Same answer
8-7 $a + b + c = 0$ *Challenge*: The only solution is $a = b = c = 0$.
8-8 $(x + y)(x - 2y - 1)$ *Challenge 1*: $(x + y + 1)(x - 2y - 3)$
8-9 36
8-10 See solution.
8-11 See solution.
8-12 No   *Challenge*: Yes
8-13 See solution.
8-14 17
8-15 f (minimum) occurs when $mx = ny$; that is, when $x:y = n:m$.

9-1 $L = 11$, $S = 5$ *Challenge*: The result is not unique, but $L = 24$, $S = 8$ seems most reasonable.

9-2 No solutions in integers   *Challenge 1*: $x = 12$, $y = 5$

9-3 9 buses, 14 cars each with 32 persons   *Challenge*: 15 buses, 21 cars each with 33 persons, 1 car with 23 persons; a second, less satisfactory, answer is 9 buses, 21 cars each with 21 persons, 1 car with 11 persons.

9-4 $1q$, $2d$, $2n$, $45c$, and $2d$, $8n$, $40c$ *Challenge*: $2d$, $8n$, $40c$

9-5 37 solutions   *Challenge*: 33 solutions

9-6 48 by 60, 40 by 72, 36 by 80

9-7 $a = n(n + 1)$, where $n = 0, 1, 2, \ldots$, or $a = m(m - 1) + 1$, where $m = 1, 2, 3, \ldots$

9-8 20 gallons

9-9 $2n - 1$ solutions, where n is the number of positive divisors of p^2.

9-10 $3A - 9$

9-11 See solution.

10-1 10   *Challenge*: 55

10-2 $f(x + 1) = f(x - 1)$ *Challenge 1*: Yes   *Challenge 2*: $1 - f(x - 1)$ *Challenge 3*: $f(x + n) = 1 - f(x - 1)$ when n is even, and $f(x + n) = f(x - 1)$ when n is odd.

10-3 (a) $ac = 1$ and $ad + b = 0$ (b) See solution.   *Challenge 2*: $ac = 1$ and $ad - b = 0$

10-4 Zero   *Challenge*: Zero

10-5 $N(\text{max}) = kV^2$, where k is a positive constant (See solution.)   *Challenge*: $N(\text{max}) = kV^{\frac{3}{2}}$

10-6 $S = \dfrac{f_3}{f_4}$ *Challenge*: $T = \dfrac{g_2{}^2 - 2g_1g_3}{g_3{}^2}$

10-7 No such values (See solution.)   *Challenge 1*: $m = 3$, $n = 2$ *Challenge 2*: $m = 5$, $n = 2$

10-8 $x = 1$ or 2   *Challenge*: $0 < x < 1$ or $x > 2$, $1 < x < 2$

10-9 Zero   *Challenge*: See solution.

10-10 $x = \sqrt{3} - 1$ *Challenge 1*: $-\sqrt{3} - 1$ *Challenge 2*: $\dfrac{\sqrt{5} + 1}{2}$

10-11 $\dfrac{1}{2\sqrt{3}}$ *Challenge 1*: $\dfrac{1}{2\sqrt{2}}$ *Challenge 2*: $\dfrac{1}{4}$ *Challenge 3*: $\dfrac{1}{2\sqrt{x}}$

10-12 a

10-13 $n \cdot 2^{n-1}$ *Challenge*: $n \cdot 3^{n-1}$, $n \cdot a^{n-1}$

10-14 $f = x^2$ *Challenge 1*: $f = (x + 1)^2$ *Challenge 2*: $f = x^2 + 1$

11-1 $P \geq 8$

11-2 $x = 32$, $y = 27$ *Challenge 1*: $x = 32$, $y = 27$
Challenge 2: $x = 45$, $y = 38$

11-3 No (See solution.)

11-4 $-1 < x < 1$ and $x > 1$ *Challenge*: $x > 1$

11-5 $(6, 6, 1)$, $(6, 5, 2)$, $(6, 4, 3)$, $(5, 5, 3)$ *Challenge*: $(7, 7, 1)$, $(7, 6, 2)$, $(7, 5, 3)$, $(6, 6, 3)$, $(6, 5, 4)$

11-6 $18\frac{1}{3}$ years   *Challenge*: $18\frac{1}{3}$ or $13\frac{1}{3}$

11-7 $A = n + 1$

11-8 R_1 is the set of real numbers, R_2 is the empty set.   *Challenge*: Same answers

11-9 See solution.   *Challenge*: See solution.

11-10 $\sqrt[10]{10!} > \sqrt[9]{9!}$

11-11 $x > 5.76$ *Challenge*: $x > 12.005$

11-12 $\frac{m}{n} = \frac{7}{5}$ (See solution.)

11-13 $4^2 = 16$ *Challenge 1*: When $a_1 = a_2 = a_3 = a_4$

12-1 $.1$ (base 9)   *Challenge 1*: $.111\ldots$ *Challenge 2*: $.14$

12-2 $x = 3 + 5k$ or $4 + 5k$, where $k = 0, \pm 1, \pm 2, \ldots$
Challenge 1: $x = 4 + 10k$ or $x = -2 + 10k$ *Challenge 2*:
$x = 5 + 17k$ or $x = -3 + 17k$

12-3 8, 2, and 4   *Challenge*: Yes

12-4 Only for $n = 1$ *Challenge*: Same answer

12-5 $y = (a, b)$; that is, y is the greatest common divisor of a and b.
Challenge: Yes

12-6 $a(x) = 1 - x^2$, $b(x) = x^2 + 4$; or $a(x) = x^2 - 1$, $b(x) = -(x^2 + 4)$

12-7 $k = 9$

12-8 $.20462046\ldots$ (base 7)   *Challenge 1*: $.25412541\ldots$
Challenge 2: $.333$

12-9 See solution.

12-10 See solution.

12-11 $x \in S$ where $S = \{0, -2, -1, 3, 6, 14\}$ *Challenge*: $x = 19$

12-12 $\sqrt{n}$

12-13 See solution.

12-14 $N = 615{,}384$ *Challenge*: $N = 820{,}512$

12-15 $N = 76$ ($N = 00$ is a trivial solution.)

12-16 $b = 6$ *Challenge*: $b = 4$

12-17 $\frac{8}{5}$ *Challenge*: $-\frac{8}{5}$

12-18 An infinite number of solutions, with $y = \pm(x^2 + 3x + 1)$ and x any integer *Challenge*: An infinite number of solutions, with $y = \pm(x^2 + 5x + 5)$ and x any integer

12-19 $(x^2 + 2x + 5)(x^2 - 8x + 20)$ *Challenge*: $(x^2 - x + 5) \times (x^2 - 9x - 4)$

12-20 $(ac + bd)^2 + (ad - bc)^2$, $(ac + bd)^2 + (bc - ad)^2$, $(ac - bd)^2 + (ad + bc)^2$, $(bd - ac)^2 + (bc + ad)^2$

12-21 8 (See solution.) *Challenge*: None

12-22 $N = 640 \ldots 0$ with n zeros, where $n = 0, 1, 2, \ldots$ *Challenge*: $N = 960 \ldots 0$

12-23 133 *Challenge 1*: 57 *Challenge 2*: See solution

12-24 $x = 3a$, where $a = 0, 1, 2, \ldots$

12-25 $A = B + C + mBC$, where m is an integer

12-26 See solution.

12-27 $R = 2$

12-28 $b \neq 4n + 3$, where $n = 0, 1, 2, \ldots$

12-29 See solution.

13-1 $r = 3$ *Challenge*: $r = \frac{1}{4}P$

13-2 a (min) $= 361$ *Challenge 1*: $C = 237$ *Challenge 2*: $C = 240 - 3k$, where $k = 1, 2, 3, \ldots$

13-3 \$1.50 *Challenge 1*: 6000 *Challenge 2*: \$9000 *Challenge 3*: \$2.00

13-4 $1:2$ *Challenge*: Same answer

13-5 $\dfrac{4}{3\sqrt{3}}$

13-6 P is 5 miles east of B.

13-7 $s = 12$ *Challenge*: $s = 1$

13-8 (a) $P_1 = L, P_2 = 0$ or $P_1 = 0, P_2 = L$ (b) $P_1 = P_2 = \dfrac{L}{2}$

13-9 $\dfrac{c^4}{2}$ *Challenge*: $\dfrac{c^3}{4}$

13-10 x equals the arithmetic mean of $k_1 + k_2 + \cdots + k_n$.

13-11 $2c + 1$

13-12 See solution.

13-13 $\dfrac{3\sqrt{3}}{4} r^2$

14-1 $x = 3$ *Challenge 1*: $x = 3$ *Challenge 2*: $x = 3$ or 8

14-2 All real values of h and k except $h = k = 0$ *Challenge*: (a) $h > k$ (b) $h < k$

14-3 $b = c = 1$ *Challenge*: $b = -2, c = 1$

14-4 $\dfrac{b^2}{ac} = \dfrac{m}{n} + 2 + \dfrac{n}{m} = \left(\sqrt{\dfrac{m}{n}} + \sqrt{\dfrac{n}{m}}\right)^2$

14-5 When $p = 20$, $x = 10 \pm 4\sqrt{5}$; when $p \neq 20$, $x = -1$ (See solution.)

14-6 $c = \dfrac{m-1}{2}$, $b = \dfrac{-m-1}{2}$

14-7 See solution. *Challenge:* $r_1 + r_2 + r_3 = -b$; $r_1r_2 + r_2r_3 + r_3r_1 = c$; $r_1r_2r_3 = -d$; $S_3 + bS_2 + cS_1 + dS_0 = 0$

14-8 $C = 90$ *Challenge 1:* $C \approx 110$ *Challenge 2:* $C = 110$

14-9 $s = \dfrac{r^2}{p^2}$, $q = \dfrac{p^2}{4} + \dfrac{2r}{p}$ *Challenge 1:* $u = v = 1$ *Challenge 2:* $v = u^2 = \dfrac{-1 - i\sqrt{3}}{2}$, where $i = \sqrt{-1}$

14-10 None (See solution.) *Challenge 1:* $n = 4$, $m = 6$ *Challenge 2:* $n = 2$, $m = 3$

14-11 $2B^3 - 9ABC + 27A^2D = 0$ *Challenge:* $B^3D - AC^3 = 0$

14-12 See solution.

14-13 See solution.

15-1 $x_1 = x_2 = \dfrac{1}{8}$, $x_3 = x_4 = \dfrac{3}{8}$

15-2 x, any real number, $y = 3x - 5a + 2c$, $z = -2x + 3a - c$, where $3a + b - c = 0$

15-3 $p^2 = 3$, $x:y = -2:1$

15-4 $a + b + c = 0$

15-5 $a:b:c = 3:-2:-1$

15-6 No finite solutions *Challenge:* $x = 3$, $y = 1$

15-7 x, any non-negative real number, y, any non-negative number, $z = \dfrac{1}{6}(b - 4x - 5y)$, where $b - 4x - 5y > 0$ and $2b = a + c$

15-8 $N_0 = 16$, $N_f = 18$, $N = 24$

16-1 Infinite *Challenge 1:* None (parallel lines) *Challenge 2:* One point $(1, 3)$

16-2 t *Challenge 1:* Yes *Challenge 2:* Yes

16-3 (a) $f = x + 2$ with $0 \leq x \leq 3$ (b) $f = -3x + 14$ with $3 \leq x \leq 4$

16-4 $m = 210$ *Challenge 1:*

$$m = a_3 \frac{c_1b_2 - c_2b_1}{a_1b_2 - a_2b_1} + b_3 \frac{a_1c_2 - a_2c_1}{a_1b_2 - a_2b_1}$$

Challenge 2: Formula of Challenge 1 inapplicable since $a_1b_2 - a_2b_1 = 0$

16-5 The non-negative portion of the y-axis   *Challenge*: The non-positive portion of the y-axis

16-6 $4X^2 - 29 = 0$

16-7 $a_1 = 2$, $a_2 = 1$ *Challenge 1*: $a_2 = 3$, $a_1 = \dfrac{2}{3}$ *Challenge 2*: $a_1 = \dfrac{(2d + 2)^2}{3d^2 + 4d + 2}$, $a_2 = \dfrac{(2c + 3)^2}{2c^2 + 4c + 3}$, where $d = -\dfrac{2c + 2}{2c + 3}$

16-8 18 miles   *Challenge*: $3\dfrac{1}{8}$ hours

16-9 $y = \dfrac{1}{2} mx^2$, $x \geq 0$ *Challenge*: $y = x(1 + \sqrt{1 + 4m^2})$, $x \geq 0$

16-10 $y = x + 3$ (See solution.)

16-11 $X > 1$ *Challenge*: $X = 1 + \sqrt{3}$ or $X = \dfrac{1 + \sqrt{5}}{2}$

16-12 See solution.   *Challenge 1*: 25 by 18   *Challenge 2*: $L' = \dfrac{1}{2} W$, $W' = 2L$

16-13 See solution.

16-14 See solution.   *Challenge 1*: See solution.   *Challenge 2*: See solution.

16-15 $S:n = 2:1$

17-1 00   *Challenge 1*: 02   *Challenge 2*: 20

17-2 $f(n + 1) = \dfrac{2f(n)f(n + 2)}{f(n) + f(n + 2)}$ with $f(1) = \dfrac{1}{4}$, $f(2) = \dfrac{1}{8}$ *Challenge 1*: Same formula with $f(1) = \dfrac{1}{3}$, $f(2) = \dfrac{1}{6}$ *Challenge 2*: Same formula with $f(1) = \dfrac{1}{2}$, $f(2) = \dfrac{1}{5}$

17-3 (a) $D = h\left(\dfrac{1 + r}{1 - r}\right)$ (b) $T = \dfrac{\sqrt{h}}{4}\left(\dfrac{1 + \sqrt{r}}{1 - \sqrt{r}}\right)$ *Challenge*: $D_A = \dfrac{\pi}{2}\left(\dfrac{1}{1 - r^2}\right)$, $D_B = \dfrac{\pi}{2}\left(\dfrac{r}{1 - r^2}\right)$

17-4 See solution.   *Challenge*: (a) a_1, a_2, a_4, a_8, and a_2, a_4, a_8, a_{16} (b) a_1, a_2, a_4, a_8, a_{16}

17-5 $d = \dfrac{12}{11}$ *Challenge 1*: $d = 12$ *Challenge 2*: $n = 3$, $d = 69$; the third term is $141 = 47n$.

17-6 n^3 *Challenge*: See solution.

17-7 Zero   *Challenge*: $n = r + s - 1$ or $n = r + s$

17-8 See solution.

17-9 $r = 2$, a is arbitrary.

17-10 $d < 10$ (degrees)   *Challenge 1*: $d < 6$ (The value obtained is $5\dfrac{5}{11}$.)   *Challenge 2*: $d < \dfrac{720}{n(n - 1)}$

17-11 $1 - \dfrac{r}{s} + \left(\dfrac{r}{s}\right)^2$

17-12 5; that is, $n \geq 5$ *Challenge 1*: $n \geq 10$ for $x = \dfrac{1}{2}$, $n \geq 3$ for $x = \dfrac{1}{8}$ *Challenge 2*: $n \geq 4$ for $x = -\dfrac{1}{4}$, $n \geq 9$ for $x = -\dfrac{1}{2}$, $n \geq 3$ for $x = -\dfrac{1}{8}$ *Challenge 3*: See solution.

17-13 $S = \dfrac{3 - \sqrt{5}}{2}$ *Challenge*: See solution.

17-14 75   *Challenge*: arbitrary

17-15 $A_1 \geq \sqrt{3}$ *Challenge 1*: $\sqrt{3}, \sqrt{3}, \sqrt{3}, \ldots$ *Challenge 2*: $2, \dfrac{9}{5}, \dfrac{7}{4}, \ldots$

17-16 $S = \dfrac{3}{4} n^2(n + 1)(5n + 1)$ *Challenge*: $S = n^2(2n^2 - 1)$

17-17 $mn = S(m + n) - S(m) - S(n)$

17-18 $a_4 = 39$

17-19 See solution.

17-20 (a) $\dfrac{1}{2} n(n^2 - n + 2j)$ (b) $\dfrac{1}{2} n(2ni - n + 1)$ (c) $\dfrac{1}{2} n(n^2 + 1)$

17-21 $P = 1 - x$, $Q = 1 + x$ *Challenge 1*: $S = \dfrac{4}{3}$ when $x = \dfrac{1}{2}$, $S = \dfrac{8}{15}$ when $x = \dfrac{1}{4}$ *Challenge 2*: $P = 1 - x$, $Q = 1 + x$

17-22 $I = \dfrac{1}{3}$

17-23 $S = \dfrac{2n}{n + 1}$ *Challenge*: $\lim\limits_{n \to \infty} S = 2$

17-24 1

17-25 $S_n = \dfrac{n}{3n + 1}$ *Challenge 1*: $S_n = \dfrac{n}{2n + 1}$ *Challenge 2*: $S_n = \dfrac{n}{n + 1}$ (See Problem 9-23.)

17-26 $T = \dfrac{(1 + r)(a_1 - a_n)}{2(1 - r)}$

17-27 See solution.

17-28 $n = 10\dfrac{11}{23}$ (days)

17-29 $S_n = 2 + (n - 1) \cdot 2^{n+1}$ *Challenge 1*: $S_n = \dfrac{1}{4}[3 + (2n - 1) 3^{n+1}]$ *Challenge 2*: $S_n = \dfrac{1}{9}[4 + (3n - 1) \cdot 4^{n+1}]$ *Challenge 3*: $S_n = \dfrac{1}{16}[5 + (4n - 1) \cdot 5^{n+1}]$

17-30 See solution.

17-31 $S_n = (n + 1)! - 1$

18-1 $x = \log_{10} 2 \approx .3010$ *Challenge:* $y = \dfrac{1}{\log_{10} 2}$

18-2 (a) $x > 0$ or $x < -2$ (b) $0 < x < 4$ *Challenge:* (a) $x > 2$ or $x < 0$ (b) $-4 < x < 0$

18-3 $f = 1 + x,\ g = 1 + y$ *Challenge:* $f = 1 - x,\ g = 1 - y$

18-4 $x = (ab)^{-1}$

18-5 $S_n = \log_N 25!$ *Challenge:* $T_n = \log_N \dfrac{(2^{12} \cdot 12!)^2}{25!}$

19-1 $\dfrac{1}{5}$ *Challenge:* $\dfrac{1}{4} + \dfrac{1}{4} = \dfrac{1}{2}$

19-2 $(.9)^4 = .6561$ *Challenge:* $.4096$

19-3 3 games out of 4   *Challenge:* 3 games out of 5

19-4 See solution.   *Challenge:* Yes

19-5 $n = 8$ *Challenge 1:* $n = 9$ *Challenge 2:* Impossible

19-6 462 (See solution.)

19-7 37-cent loss

19-8 $N = r + rs + rst$

19-9 16

19-10 See solution.

19-11 30

19-12 1092 (See solution.)

19-13 $n_1 = 462,\ n_2 = 252$

20-1 $(2n + 1)!! = \dfrac{(2n + 1)!}{2^n \cdot n!}$

20-2 8   *Challenge 1:* No change   *Challenge 2:* No change

20-3 $\dfrac{2}{3} L$ *Challenge:* $\dfrac{3}{4} L$

20-4 The empty set   *Challenge 1:* $k = 4, n = 5$; $k = 6, n = 10$; $k = 8, n = 17$ *Challenge 2:* None

20-5 $\dfrac{1}{3}$

20-6 $F = \dfrac{1}{x^3} + \dfrac{2}{x^2} + \dfrac{3}{x} + \dfrac{4}{x - 1}$ *Challenge:* $F = -\dfrac{1}{x^3} - \dfrac{1}{x} + \dfrac{8}{x + 1}$

20-7 33

20-8 $a^2 = b + 1$ *Challenge:* No solution possible

20-9 $R = \sqrt{6}$ *Challenge 1:* $\sqrt{-2}$ *Challenge 2:* $2\sqrt{c}, 2\sqrt{-d}$, where $c = \dfrac{\sqrt{a^2 + b} + a}{2}$, $d = \dfrac{\sqrt{a^2 + b} - a}{2}$

20-10 $n = 4k$ or $n + 1 = 4k$, where $k = 1, 2, 3, \ldots$

20-11 $S = 17$

20-12 See solution.

20-13 Row 46, column 1; row 21, column 2

20-14 $P(c) = 220c^2 - 76c + 244$ $\qquad$ *Challenge*: $P(c) = 170c^2 - 76c + 244$

20-15 $r_n = b_0 + b_1 + b_2 + \cdots + b_n$ $\quad$ *Challenge*: $\frac{1}{2}(n+1)(n+2)$

20-16 $P = \frac{2}{3}$

20-17 $F = \dfrac{-23\sqrt[3]{2} + 31\sqrt[3]{4} + 12}{471}$ $\qquad$ *Challenge*: $\dfrac{\sqrt[3]{4} - \sqrt[3]{2}}{1} = \sqrt[3]{4} - \sqrt[3]{2}$

20-18 See solution.

20-19 $a_n = 1$ when $n = \frac{1}{2}k^2 + \frac{3}{2}k$, where $k = 1, 2, 3, \ldots$; $a_n = 0$ otherwise $\quad$ *Challenge*: $a_n = 1$ when $n = \frac{1}{2}k^2 + \frac{1}{2}k$, where $k = 1, 2, 3, \ldots$; $a_n = 0$ otherwise

APPENDICES

APPENDIX I   Terminating Digits

For any natural number n, n^5 TD n; that is, the terminal digit of n^5 is the same as that of n itself.

Proof

(1) Certainly it is true that 1^5 TD 1.

(2) Assume that k is the largest value of n for which the theorem is true; that is, assume that k^5 TD k.

Then $(k + 1)^5 = k^5 + 5k^4 + 10k^3 + 10k^2 + 5k + 1$.

Whether k is even or odd, the terminating digit of $10k^3 + 10k^2$ is 0. Now consider $5k^4 + 5k = 5k(k^3 + 1)$. If k is even, then the terminating digit of $5k(k^3 + 1)$ is zero. If k is odd, then k^3 is odd and $k^3 + 1$ is even. Therefore, the terminating digit of $5k(k^3 + 1)$ is 0, whether k is even or odd. It follows that the terminating digit of $(k + 1)^5$ is the same as that of $k^5 + 1$. But, by assumption k^5 TD k. Therefore, $(k + 1)^5$ TD $k + 1$.

(3) We now know that, no matter what the value of k, the theorem is true for the successor of k; that is, the theorem is true for all natural numbers n, since it is true for $k = 1$.

It can also be shown that n^{4m+1} TD n, when $m = 0, 1, 2, \ldots$.

APPENDIX II   Remainder and Factor Theorems

If a polynominal $P(x) = x^n + c_1 x^{n-1} + \cdots + c_{n-1}x + c_n$ is divided by $x - a$ until no x appears in the remainder, the remainder has the value $P(a) = a^n + c_1 a^{n-1} + \cdots + c_n$.

For example, if $P(x) = x^3 - 2x^2 + 3$ is divided by $x - 1$, the remainder is $P(1) = 1 - 2 + 3 = 2$.

Proof

$P(x) = Q(x)(x - a) + R$, by the definition of division; R is a constant. We have, upon substituting a for x, $P(a) = 0 + R$; that is, the remainder equals $P(a)$.

If $R = 0$, the division is exact and $x - a$ is a factor of $P(x)$. Conversely, if $x - a$ is a factor of $P(x)$, then $R = 0$, and the division is exact.

APPENDIX III   Maximum Product, Minimum Sum

Let $S = a + b$, where S is a constant and a and b are positive numbers. Let $P = ab$. Then $P(\text{maximum}) = \left(\dfrac{a + b}{2}\right)^2 = \dfrac{S^2}{4}$.

Proof

Since $S = a + b$, $b = S - a$. $\therefore P = a(S - a) = Sa - a^2$.

$P = \dfrac{S^2}{4} - \left(a^2 - Sa + \dfrac{S^2}{4}\right) = \dfrac{S^2}{4} - \left(a - \dfrac{S}{2}\right)^2$. The largest value of P, namely $\dfrac{S^2}{4}$, occurs when $a = \dfrac{S}{2}$, and when $a = \dfrac{S}{2}$, then $b = \dfrac{S}{2}$.

Therefore, $P(\text{max}) = \dfrac{S^2}{4} = \left(\dfrac{a + b}{2}\right)^2$.

Let $P = ab$ where P is a constant and a and b are positive numbers. Let $S = a + b$. Then $S(\text{minimum}) = 2\sqrt{ab} = 2\sqrt{P}$.

Proof

$(\sqrt{a} - \sqrt{b})^2 \geq 0$, $\therefore a + b \geq 2\sqrt{ab}$. The equality sign holds only when $\sqrt{a} = \sqrt{b}$, and when $\sqrt{a} = \sqrt{b}$, then $a = b$. Therefore, $S = a + b = 2\sqrt{P}$ is a minimum when $P = a^2 = b^2$, that is, when $a = b = \sqrt{P} = \sqrt{ab}$.

APPENDIX IV   Means

For positive numbers, the harmonic mean (H.M.) $\leq$ the geometric mean (G.M.) $\leq$ the arithmetic mean (A.M.), where

$$\text{H.M.} = \left(\frac{a_1^{-1} + a_2^{-1} + \cdots + a_n^{-1}}{n}\right)^{-1} = \frac{n}{\dfrac{1}{a_1} + \dfrac{1}{a_2} + \cdots + \dfrac{1}{a_n}},$$

$$\text{G.M.} = \sqrt[n]{a_1 a_2 \cdots a_n},$$

$$\text{A.M.} = \frac{a_1 + a_2 + \cdots + a_n}{n}.$$

Proof

(1) Let $g = \sqrt[n]{a_1 a_2 \cdots a_n}$; then $1 = \sqrt[n]{\dfrac{a_1}{g} \dfrac{a_2}{g} \cdots \dfrac{a_n}{g}}$, and $1 = \dfrac{a_1}{g} \dfrac{a_2}{g} \cdots \dfrac{a_n}{g}$. But $\dfrac{a_1}{g} + \dfrac{a_2}{g} + \cdots + \dfrac{a_n}{g} \geq n$ (see Lemma below). Therefore $\dfrac{a_1 + a_2 + \cdots + a_n}{n} \geq g$; that is, G.M. $\leq$ A.M.

(2) Using this result, we have $\sqrt[n]{a_1{}^\alpha a_2{}^\alpha \cdots a_n{}^\alpha} \leq \dfrac{a_1{}^\alpha + a_2{}^\alpha + \cdots + a_n{}^\alpha}{n}$. When $\dfrac{1}{\alpha} < 0$, $[\sqrt[n]{a_1{}^\alpha a_2{}^\alpha \cdots a_n{}^\alpha}]^{\frac{1}{\alpha}} \geq \left[\dfrac{a_1{}^\alpha + a_2{}^\alpha + \cdots + a_n{}^\alpha}{n}\right]^{\frac{1}{\alpha}}$. Take $\alpha = -1$; then $\sqrt[n]{a_1 a_2 \cdots a_n} \geq \left[\dfrac{a_1{}^{-1} + a_2{}^{-1} + \cdots + a_n{}^{-1}}{n}\right]^{-1}$, so that H.M. $\leq$ G.M.

LEMMA: If the product of n positive numbers equals 1, their sum is not less than n. The proof by Mathematical Induction follows.

(1) The theorem is true for $n = 2$. Since $(a_1 - a_2)^2 \geq 0, a_1{}^2 + a_2{}^2 \geq 2a_1 a_2$. Therefore, $\dfrac{a_1}{a_2} + \dfrac{a_2}{a_1} \geq 2$, since a positive number plus its reciprocal ≥ 2. Because $a_1 a_2 = 1, a_2 = \dfrac{1}{a_1}$, and therefore, $\dfrac{a_1}{a_2} + \dfrac{a_2}{a_1} = a_1{}^2 + \dfrac{1}{a_1{}^2} \geq 2$.

(2) Assume $a_1 + a_2 + \cdots + a_k \geq k$ when $a_1 a_2 \cdots a_k = 1$.

CASE I: If $a_1 = a_2 = \cdots = a_k = 1$, then $a_1 + a_2 + \cdots + a_k + a_{k+1} = k + 1$.

CASE II: Some of the numbers are greater than 1, and some smaller; say $a_1 < 1, a_{k+1} > 1$. Then $b a_2 a_3 \cdots a_k = 1$ where $b = a_1 a_{k+1}$. Therefore $b + a_2 + a_3 + \cdots + a_k \geq k$. But $a_1 + a_2 + \cdots + a_k + a_{k+1} = (b + a_2 + a_3 + \cdots + a_k) + a_{k+1} - b + a_1 \geq k + a_{k+1} - b + a_1 = k + 1 + a_{k+1} - b + a_1 - 1$.

Therefore $a_1 + a_2 + \cdots + a_{k+1} \geq (k + 1) + a_{k+1} - a_1 a_{k+1} + a_1 - 1 = (k + 1) + a_{k+1}(1 - a_1) - (1 - a_1) = (k + 1) + (a_{k+1} - 1)(1 - a_1)$.

Since $a_1 < 1$ and $a_{k+1} > 1$, $(a_{k+1} - 1)(1 - a_1) > 0$.

Therefore $a_1 + a_2 + \cdots + a_k + a_{k+1} \geq k + 1 +$ a positive number $> k + 1$.

APPENDIX V   Divisibility

Let $N = a_n a_{n-1} \cdots a_1 a_0$ be an integer with $n + 1$ digits, expressed in base 10. Then divisibility can be determined by the following theorems:

(a) 2 divides N if 2 divides a_0. (i.e. N is exactly divisible by 2 if A_0 is exactly divisible by 2.)

(b) 5 divides N if $a_0 = 0$ or $a_0 = 5$, because multiples of 5 end either in 5 or 0.

(c) 3 divides N if 3 divides S where $S = a_n + a_{n-1} + \cdots + a_1 + a_0$.

9 divides N if 9 divides S where $S = a_n + a_{n-1} + \cdots + a_1 + a_0$.

(d) 7 divides N if 7 divides P where $P = (1 \cdot a_0 + 3a_1 + 2a_2) - (1 \cdot a_3 + 3a_4 + 2a_5) + (1 \cdot a_6 + 3a_7 + 2a_8) - \cdots$

(e) 11 divides N if 11 divides Q where $Q = a_0 - a_1 + a_2 - \cdots \pm a_n$.

Proofs

(c) Let $N = a_8 a_7 a_6 a_5 a_4 a_3 a_2 a_1 a_0$. (The proof of the general case $a_n a_{n-1} \cdots a_1 a_0$ is similar.)

$N = a_8(9 + 1)^8 + a_7(9 + 1)^7 + \cdots + a_1(9 + 1) + a_0.$

Using the expression $M_i(9)$ to mean a multiple of 9, for $i = 1, 2, 3, 4, 5, 6, 7, 8$, we can rewrite N as $N = a_8[M_8(9) + 1] + a_7[M_7(9) + 1] + \cdots + a_1[M_1(9) + 1] + a_0.$ (See Appendix VI.)

Therefore $N = M(9) + a_8 + a_7 + a_6 + a_5 + a_4 + a_3 + a_2 + a_1 + a_0.$ Certainly a multiple of 9, $M(9)$, is exactly divisible by 3. So whenever $S = a_8 + a_7 + \cdots + a_1 + a_0$ is exactly divisible by 3, then 3 divides N. And whenever S is divisible by 9, then 9 divides N.

(d) Taking $N = a_8 a_7 a_6 \cdots a_1 a_0.$

$N = a_8(7 + 3)^8 + a_7(7 + 3)^7 + \cdots + a_1(7 + 3) + a_0.$

$N = a_8[M_8(7) + 3^8] + a_7[M_7(7) + 3^7] + \cdots + a_1[M_1(7) + 3] + a_0.$ (See Appendix VI.)

$N = M(7) + 3^8 a_8 + 3^7 a_7 + \cdots + 3a_1 + a_0.$

Since $3^2 = 7 + 2, 3^3 = 4 \cdot 7 - 1, 3^4 = 12 \cdot 7 - 3, 3^5 = 35 \cdot 7 - 2, 3^6 = 104 \cdot 7 + 1, 3^7 = 312 \cdot 7 + 3, 3^8 = 937 \cdot 7 + 2,$

$N = M^*(7) + (a_0 + 3a_1 + 2a_2) - (a_3 + 3a_4 + 2a_5) + (a_6 + 3a_7 + 2a_8).$

Since a proof of this kind can be applied to the general case when $N = a_n a_{n-1} \cdots a_1 a_0$, whenever 7 divides P, 7 divides N.

(e) Let $N = a_8 a_7 a_6 \cdots a_1 a_0.$

$N = a_8(11 - 1)^8 + a_7(11 - 1)^7 + \cdots + a_1(11 - 1) + a_0.$

$N = a_8[M_8(11) + 1] + a_7[M_7(11) - 1] + \cdots + a_1[M_1(11) - 1] + a_0.$ (See Appendix VI.)

$N = M(11) + a_8 - a_7 + a_6 - a_5 + a_4 - a_3 + a_2 - a_1 + a_0.$

Therefore, if 11 divides Q, 11 divides N.

ILLUSTRATION 1: We can test $N = 317,142$ for divisibility by 7 using Theorem (d) or modular division.

(1) By Theorem (d), we have

$(1 \cdot 2 + 3 \cdot 4 + 2 \cdot 1) - (1 \cdot 7 + 3 \cdot 1 + 2 \cdot 3) = 16 - 16 = 0.$

Since 0 is divisible by 7, 7 divides 317,142.

(2) Using modular division, we have

$N = 3 \cdot 10^5 + 1 \cdot 10^4 + 7 \cdot 10^3 + 1 \cdot 10^2 + 4 \cdot 10 + 2.$

We use synthetic division, reducing coefficients modulo 7 as we proceed with the division. The divisor is $10 - 7$, or 3.

3	1	$\not{7}0$	$\not{1}$	4	2	$\lfloor 3$
	$\not{9}2$	$\not{9}2$	6	0	$\not{1}\not{2}5$	
3	3	2	$\not{7}0$	4	$\not{7}0$	

Since the remainder is 0, the division is exact, and 7 must divide N.

ILLUSTRATION 2: Test $N = 41,631$ for divisibility by 7, by Theorem (d), and by modular division.

(1) By Theorem (d)

$(1 \cdot 1 + 3 \cdot 3 + 2 \cdot 6) - (1 \cdot 1 + 3 \cdot 4) = 22 - 13 = 9.$

Since 9 divided by 7 leaves a remainder of 2, 7 does not divide N.

(2) By modular division, (reducing coefficients modulo 7)

4	1	6	3	1	$\lfloor 3$
	$\not{1}\not{2}5$	$\not{1}\not{8}4$	$\not{9}2$	$\not{1}\not{5}1$	
4	6	$\not{1}\not{2}3$	5	2	

Since the remainder is 2, the division is not exact and 7 does not divide N.

ILLUSTRATION 3: Test $N = 317,152$ for divisibility by 11, by Theorem (e), and by modular division.

(1) By Theorem (e)

$(2 + 1 + 1) - (5 + 7 + 3) = 4 - 15 = -11.$

Therefore, 11 must divide N.

(2) The divisor for modular division is $10 - 11$, or -1.

3	1	7	1	5	2	$\lfloor -1$
	-3	2	-9	8	-2	
3	-2	9	-8	̶1̶3̶2	0	

Since the remainder is 0, 11 must divide N.

ILLUSTRATION 4: Test $N = 71,351$ for divisibility by 11, by Theorem (e), and by modular division.

(1) Theorem (e) shows that 11 does not divide N.

$$(1 + 3 + 7) - (5 + 1) = 11 - 6 = 5$$

(2) Modular division shows the same result.
The divisor is $10 - 11$, or -1.

7	1	3	5	1	$\lfloor -1$
	-7	6	-9	4	
7	-6	9	-4	5	

ILLUSTRATION 5: Test $N = 24,041$ for divisibility by 13, using modular division. The divisor is $10 - 13$, or -3.

2	4	0	4	1	$\lfloor -3$
	-6	6	$-1̶8 - 5$	3	
2	-2	6	-1	4	

Since the remainder is 4, the division is not exact and 13 does not divide 24,041.

GENERAL THEOREM: A number N, written in base b, is divisible by $b + 1$ if the sum of the odd-numbered digits, less the sum of the even-numbered digits, is divisible by $b + 1$. The absolute value of the difference of the sums is used.

Specific Case 1: A number N, written in base 10, is divisible by 11_{10} if the sum of the even-numbered digits less the sum of the odd-numbered digits is divisible by 11_{10}.

Specific Case 2: A number N, written in base 9, is divisible by 11_9 if the sum of the even-numbered digits less the sum of the odd-numbered digits is divisible by 11_9.

Example 1: 52789_{10} is divisible by 11_{10} since $5 + 7 + 9 = 21_{10}$ less $2 + 8 = 10_{10}$ equals 11_{10}, which is obviously divisible by 11_{10}. Actually, $52789_{10} \div 11_{10} = 4799_{10}$.

Example 2: 718181_9 is divisible by 11_9 since $7 + 8 + 8 = 25_9$ less $1 + 1 + 1 = 3_9$ equals 22_9, which is divisible by 11_{10}. Actually, $718181_9 \div 11_9 = 64371_9$.

Example 3: 18192_{12} is divisible by 11_{12} since $8 + 9 = 15_{12}$ less $1 + 1 + 2 = 4_{12}$ equals 11_{12}, which is obviously divisible by 11_{12}. Actually, $18192_{12} \div 11_{12} = 16720_{12}$.

Proof

$$N = a_0b^n + a_1b^{n-1} + a_2b^{n-2} + \cdots + a_{n-2}b^2 + a_{n-1}b + a_n$$
$$\therefore N = a_0[(b + 1) - 1]^n + a_1[(b + 1) - 1]^{n-1} + \cdots + a_{n-1}[(b + 1) - 1] + a_n$$

After expansion we can express N as $M_1(b + 1) + M_2(b + 1) + \cdots + M_{n-1}(b + 1) + R$, where $M_i(b + 1)$ represents a multiple of $b + 1$, and

$$R = a_0 - a_1 + a_2 - \cdots - a_{n-1} + a_n, \text{ if } n \text{ is even, and}$$
$$R = -a_0 + a_1 - a_2 + \cdots - a_n + a_n, \text{ if } n \text{ is odd.}$$

$\therefore N = M(b + 1) + R$, where $M(b + 1) = M_1(b + 1) + M_2(b + 1) + \cdots + M_{n-1}(b + 1)$.
$\therefore N$ is divisible by $b + 1$, if R is divisible by $b + 1$.

APPENDIX VI   Binomial Theorem

If a and b are two real numbers and n is a natural number, then

$$(a + b)^n = \binom{n}{0} a^nb^0 + \binom{n}{1} a^{n-1}b^1 + \binom{n}{2} a^{n-2}b^2 + \cdots + \binom{n}{k} a^{n-k}b^k + \cdots + \binom{n}{n-1} a^1b^{n-1} + \binom{n}{n} a^0b^n,$$

where $\binom{n}{k} = \dfrac{n!}{k!(n - k)!}$.

The number of terms in the expansion of $(a + b)^n$ is $n + 1$. Two terms symmetrically placed with respect to the beginning and end of the expansion have equal coefficients. That is,

$$\binom{n}{k} = \binom{n}{n - k}.$$

ILLUSTRATION: $(a + b)^4 = a^4 + 4a^3b + 6a^2b^2 + 4ab^3 + b^4$

APPENDIX VII   Miscellaneous

1. Factors

(a) Factors of the sum of two cubes

$$x^3 + y^3 = (x + y)(x^2 - xy + y^2)$$

(b) Factors of the difference of two cubes

$$x^3 - y^3 = (x - y)(x^2 + xy + y^2)$$

2. Summations

(a)
$$S = 1 + 2 + 3 + \cdots + n = \frac{1}{2} n(n + 1)$$
$$S = 1 + 3 + 5 + \cdots + 2n - 1 = n^2$$
$$S = 1^2 + 2^2 + 3^2 + \cdots + n^2 = \left(\frac{1}{6} n\right)(n + 1)(2n + 1)$$
$$S = 1^3 + 2^3 + 3^3 + \cdots + n^3 = \frac{1}{4} n^2(n + 1)^2$$

(b)
$$S = a + (a + d) + (a + 2d) + \cdots + [a + (n - 1)d] = \frac{1}{2} n[2a + (n - 1)d]$$

$$S = a + ar + ar^2 + \cdots + ar^{n-1} = \frac{a - ar^n}{1 - r}, r \neq 1$$

$$S = a + ar + ar^2 + \cdots = \frac{a}{1 - r}, |r| < 1.$$

3. Cramer's Rule

The solutions of the system of the linear equations

$$a_1x + b_1y + c_1z = d_1$$
$$a_2x + b_2y + c_2z = d_2$$
$$a_3x + b_3y + c_3z = d_3$$

are $x = \dfrac{D_x}{D}$, $y = \dfrac{D_y}{D}$, $z = \dfrac{D_z}{D}$ where $D = \begin{vmatrix} a_1 & b_1 & c_1 \\ a_2 & b_2 & c_2 \\ a_3 & b_3 & c_3 \end{vmatrix} = $

$$a_1b_2c_3 + a_2b_3c_1 + a_3b_1c_2 - a_1b_3c_2 - a_2b_1c_3 - a_3b_2c_1 \neq 0$$

$$D_x = \begin{vmatrix} d_1 & b_1 & c_1 \\ d_2 & b_2 & c_2 \\ d_3 & b_3 & c_3 \end{vmatrix}, \quad D_y = \begin{vmatrix} a_1 & d_1 & c_1 \\ a_2 & d_2 & c_2 \\ a_3 & d_3 & c_3 \end{vmatrix}, \quad D_z = \begin{vmatrix} a_1 & b_1 & d_1 \\ a_2 & b_2 & d_2 \\ a_3 & b_3 & d_3 \end{vmatrix}.$$

4. Polygonal Numbers

We designate the nth r-agonal number by $P_n{}^r$, where $n = 1, 2, 3, \ldots$ and $r = 2, 3, 4, \ldots$.

(a) $P_n{}^2$, the nth linear number is $n = n + \dfrac{0 \cdot n(n-1)}{2}$.

(b) $P_n{}^3$, the nth triangular number is $\dfrac{n(n+1)}{2} = n + \dfrac{1 \cdot n(n-1)}{2}$.

(c) $P_n{}^4$, the nth square number is $n^2 = n + \dfrac{2n(n-1)}{2}$.

(d) $P_n{}^5$, the nth pentagonal number is $\dfrac{n(3n-1)}{2} = n + \dfrac{3n(n-1)}{2}$.

(e) $P_n{}^6$, the nth hexagonal number is $\dfrac{n(4n-2)}{2} = n + \dfrac{4n(n-1)}{2}$.

(f) $P_n{}^r$, the nth r-agonal number is

$$\frac{n[2 + (n-1)(r-2)]}{2} = n + \frac{n(r-2)(n-1)}{2}.$$

5. Finite Mathematical Induction

To prove that a formula or sentence involving one or more variables is true for all natural numbers greater than or equal to a given natural number, it is sufficient to show that

(a) the formula holds for the given natural number (usually the number 1),

(b) if the formula holds for the natural number k, where $k \geq 1$, (or the given natural number), it also holds for $k + 1$.

Step (a) is referred to as the verification step.

Step (b) is referred to as the induction step.

A CATALOG OF SELECTED
DOVER BOOKS
IN ALL FIELDS OF INTEREST

FRANK LLOYD WRIGHT'S HOLLYHOCK HOUSE, Donald Hoffmann. Lavishly illustrated, carefully documented study of one of Wright's most controversial residential designs. Over 120 photographs, floor plans, elevations, etc. Detailed perceptive text by noted Wright scholar. Index. 128pp. 9¼ × 10¾.
27133-1 Pa. $11.95

THE MALE AND FEMALE FIGURE IN MOTION: 60 Classic Photographic Sequences, Eadweard Muybridge. 60 true-action photographs of men and women walking, running, climbing, bending, turning, etc., reproduced from rare 19th-century masterpiece. vi + 121pp. 9 × 12.
24745-7 Pa. $10.95

1001 QUESTIONS ANSWERED ABOUT THE SEASHORE, N. J. Berrill and Jacquelyn Berrill. Queries answered about dolphins, sea snails, sponges, starfish, fishes, shore birds, many others. Covers appearance, breeding, growth, feeding, much more. 305pp. 5¼ × 8¼.
23366-9 Pa. $7.95

GUIDE TO OWL WATCHING IN NORTH AMERICA, Donald S. Heintzelman. Superb guide offers complete data and descriptions of 19 species: barn owl, screech owl, snowy owl, many more. Expert coverage of owl-watching equipment, conservation, migrations and invasions, etc. Guide to observing sites. 84 illustrations. xiii + 193pp. 5⅜ × 8½.
27344-X Pa. $7.95

MEDICINAL AND OTHER USES OF NORTH AMERICAN PLANTS: A Historical Survey with Special Reference to the Eastern Indian Tribes, Charlotte Erichsen-Brown. Chronological historical citations document 500 years of usage of plants, trees, shrubs native to eastern Canada, northeastern U.S. Also complete identifying information. 343 illustrations. 544pp. 6½ × 9¼.
25951-X Pa. $12.95

STORYBOOK MAZES, Dave Phillips. 23 stories and mazes on two-page spreads: Wizard of Oz, Treasure Island, Robin Hood, etc. Solutions. 64pp. 8¼ × 11.
23628-5 Pa. $2.95

NEGRO FOLK MUSIC, U.S.A., Harold Courlander. Noted folklorist's scholarly yet readable analysis of rich and varied musical tradition. Includes authentic versions of over 40 folk songs. Valuable bibliography and discography. xi + 324pp. 5⅜ × 8½.
27350-4 Pa. $7.95

MOVIE-STAR PORTRAITS OF THE FORTIES, John Kobal (ed.). 163 glamor, studio photos of 106 stars of the 1940s: Rita Hayworth, Ava Gardner, Marlon Brando, Clark Gable, many more. 176pp. 8⅜ × 11¼.
23546-7 Pa. $10.95

BENCHLEY LOST AND FOUND, Robert Benchley. Finest humor from early 30s, about pet peeves, child psychologists, post office and others. Mostly unavailable elsewhere. 73 illustrations by Peter Arno and others. 183pp. 5⅜ × 8½.
22410-4 Pa. $5.95

YEKL and THE IMPORTED BRIDEGROOM AND OTHER STORIES OF YIDDISH NEW YORK, Abraham Cahan. Film Hester Street based on Yekl (1896). Novel, other stories among first about Jewish immigrants on N.Y.'s East Side. 240pp. 5⅜ × 8½.
22427-9 Pa. $6.95

SELECTED POEMS, Walt Whitman. Generous sampling from *Leaves of Grass*. Twenty-four poems include "I Hear America Singing," "Song of the Open Road," "I Sing the Body Electric," "When Lilacs Last in the Dooryard Bloom'd," "O Captain! My Captain!"—all reprinted from an authoritative edition. Lists of titles and first lines. 128pp. 5³⁄₁₆ × 8¼.
26878-0 Pa. $1.00

THE BEST TALES OF HOFFMANN, E. T. A. Hoffmann. 10 of Hoffmann's most important stories: "Nutcracker and the King of Mice," "The Golden Flowerpot," etc. 458pp. 5⅜ × 8½. 21793-0 Pa. $8.95

FROM FETISH TO GOD IN ANCIENT EGYPT, E. A. Wallis Budge. Rich detailed survey of Egyptian conception of "God" and gods, magic, cult of animals, Osiris, more. Also, superb English translations of hymns and legends. 240 illustrations. 545pp. 5⅜ × 8½. 25803-3 Pa. $11.95

FRENCH STORIES/CONTES FRANÇAIS: A Dual-Language Book, Wallace Fowlie. Ten stories by French masters, Voltaire to Camus: "Micromegas" by Voltaire; "The Atheist's Mass" by Balzac; "Minuet" by de Maupassant; "The Guest" by Camus, six more. Excellent English translations on facing pages. Also French-English vocabulary list, exercises, more. 352pp. 5⅜ × 8½. 26443-2 Pa. $8.95

CHICAGO AT THE TURN OF THE CENTURY IN PHOTOGRAPHS: 122 Historic Views from the Collections of the Chicago Historical Society, Larry A. Viskochil. Rare large-format prints offer detailed views of City Hall, State Street, the Loop, Hull House, Union Station, many other landmarks, circa 1904–1913. Introduction. Captions. Maps. 144pp. 9⅜ × 12¼. 24656-6 Pa. $12.95

OLD BROOKLYN IN EARLY PHOTOGRAPHS, 1865–1929, William Lee Younger. Luna Park, Gravesend race track, construction of Grand Army Plaza, moving of Hotel Brighton, etc. 157 previously unpublished photographs. 165pp. 8⅞ × 11¾. 23587-4 Pa. $13.95

THE MYTHS OF THE NORTH AMERICAN INDIANS, Lewis Spence. Rich anthology of the myths and legends of the Algonquins, Iroquois, Pawnees and Sioux, prefaced by an extensive historical and ethnological commentary. 36 illustrations. 480pp. 5⅜ × 8½. 25967-6 Pa. $8.95

AN ENCYCLOPEDIA OF BATTLES: Accounts of Over 1,560 Battles from 1479 B.C. to the Present, David Eggenberger. Essential details of every major battle in recorded history from the first battle of Megiddo in 1479 B.C. to Grenada in 1984. List of Battle Maps. New Appendix covering the years 1967–1984. Index. 99 illustrations. 544pp. 6½ × 9¼. 24913-1 Pa. $14.95

SAILING ALONE AROUND THE WORLD, Captain Joshua Slocum. First man to sail around the world, alone, in small boat. One of great feats of seamanship told in delightful manner. 67 illustrations. 294pp. 5⅜ × 8½. 20326-3 Pa. $5.95

ANARCHISM AND OTHER ESSAYS, Emma Goldman. Powerful, penetrating, prophetic essays on direct action, role of minorities, prison reform, puritan hypocrisy, violence, etc. 271pp. 5⅜ × 8½. 22484-8 Pa. $5.95

MYTHS OF THE HINDUS AND BUDDHISTS, Ananda K. Coomaraswamy and Sister Nivedita. Great stories of the epics; deeds of Krishna, Shiva, taken from puranas, Vedas, folk tales; etc. 32 illustrations. 400pp. 5⅜ × 8½. 21759-0 Pa. $9.95

BEYOND PSYCHOLOGY, Otto Rank. Fear of death, desire of immortality, nature of sexuality, social organization, creativity, according to Rankian system. 291pp. 5⅜ × 8½. 20485-5 Pa. $7.95

A THEOLOGICO-POLITICAL TREATISE, Benedict Spinoza. Also contains unfinished Political Treatise. Great classic on religious liberty, theory of government on common consent. R. Elwes translation. Total of 421pp. 5⅜ × 8½. 20249-6 Pa. $8.95

MY BONDAGE AND MY FREEDOM, Frederick Douglass. Born a slave, Douglass became outspoken force in antislavery movement. The best of Douglass' autobiographies. Graphic description of slave life. 464pp. 5⅜ × 8½.   22457-0 Pa. $8.95

FOLLOWING THE EQUATOR: A Journey Around the World, Mark Twain. Fascinating humorous account of 1897 voyage to Hawaii, Australia, India, New Zealand, etc. Ironic, bemused reports on peoples, customs, climate, flora and fauna, politics, much more. 197 illustrations. 720pp. 5⅜ × 8½.   26113-1 Pa. $15.95

THE PEOPLE CALLED SHAKERS, Edward D. Andrews. Definitive study of Shakers: origins, beliefs, practices, dances, social organization, furniture and crafts, etc. 33 illustrations. 351pp. 5⅜ × 8½.   21081-2 Pa. $8.95

THE MYTHS OF GREECE AND ROME, H. A. Guerber. A classic of mythology, generously illustrated, long prized for its simple, graphic, accurate retelling of the principal myths of Greece and Rome, and for its commentary on their origins and significance. With 64 illustrations by Michelangelo, Raphael, Titian, Rubens, Canova, Bernini and others. 480pp. 5⅜ × 8½.   27584-1 Pa. $9.95

PSYCHOLOGY OF MUSIC, Carl E. Seashore. Classic work discusses music as a medium from psychological viewpoint. Clear treatment of physical acoustics, auditory apparatus, sound perception, development of musical skills, nature of musical feeling, host of other topics. 88 figures. 408pp. 5⅜ × 8½.   21851-1 Pa. $9.95

THE PHILOSOPHY OF HISTORY, Georg W. Hegel. Great classic of Western thought develops concept that history is not chance but rational process, the evolution of freedom. 457pp. 5⅜ × 8½.   20112-0 Pa. $9.95

THE BOOK OF TEA, Kakuzo Okakura. Minor classic of the Orient: entertaining, charming explanation, interpretation of traditional Japanese culture in terms of tea ceremony. 94pp. 5⅜ × 8½.   20070-1 Pa. $2.95

LIFE IN ANCIENT EGYPT, Adolf Erman. Fullest, most thorough, detailed older account with much not in more recent books, domestic life, religion, magic, medicine, commerce, much more. Many illustrations reproduce tomb paintings, carvings, hieroglyphs, etc. 597pp. 5⅜ × 8½.   22632-8 Pa. $10.95

SUNDIALS, Their Theory and Construction, Albert Waugh. Far and away the best, most thorough coverage of ideas, mathematics concerned, types, construction, adjusting anywhere. Simple, nontechnical treatment allows even children to build several of these dials. Over 100 illustrations. 230pp. 5⅜ × 8½.   22947-5 Pa. $7.95

DYNAMICS OF FLUIDS IN POROUS MEDIA, Jacob Bear. For advanced students of ground water hydrology, soil mechanics and physics, drainage and irrigation engineering, and more. 335 illustrations. Exercises, with answers. 784pp. 6⅛ × 9¼.   65675-6 Pa. $19.95

SONGS OF EXPERIENCE: Facsimile Reproduction with 26 Plates in Full Color, William Blake. 26 full-color plates from a rare 1826 edition. Includes "The Tyger," "London," "Holy Thursday," and other poems. Printed text of poems. 48pp. 5¼ × 7.   24636-1 Pa. $4.95

OLD-TIME VIGNETTES IN FULL COLOR, Carol Belanger Grafton (ed.). Over 390 charming, often sentimental illustrations, selected from archives of Victorian graphics—pretty women posing, children playing, food, flowers, kittens and puppies, smiling cherubs, birds and butterflies, much more. All copyright-free. 48pp. 9¼ × 12¼.   27269-9 Pa. $5.95

PERSPECTIVE FOR ARTISTS, Rex Vicat Cole. Depth, perspective of sky and sea, shadows, much more, not usually covered. 391 diagrams, 81 reproductions of drawings and paintings. 279pp. 5⅜ × 8½. 22487-2 Pa. $6.95

DRAWING THE LIVING FIGURE, Joseph Sheppard. Innovative approach to artistic anatomy focuses on specifics of surface anatomy, rather than muscles and bones. Over 170 drawings of live models in front, back and side views, and in widely varying poses. Accompanying diagrams. 177 illustrations. Introduction. Index. 144pp. 8⅜ × 11¼. 26723-7 Pa. $7.95

GOTHIC AND OLD ENGLISH ALPHABETS: 100 Complete Fonts, Dan X. Solo. Add power, elegance to posters, signs, other graphics with 100 stunning copyright-free alphabets: Blackstone, Dolbey, Germania, 97 more—including many lower-case, numerals, punctuation marks. 104pp. 8⅛ × 11. 24695-7 Pa. $7.95

HOW TO DO BEADWORK, Mary White. Fundamental book on craft from simple projects to five-bead chains and woven works. 106 illustrations. 142pp. 5⅜ × 8. 20697-1 Pa. $4.95

THE BOOK OF WOOD CARVING, Charles Marshall Sayers. Finest book for beginners discusses fundamentals and offers 34 designs. "Absolutely first rate . . . well thought out and well executed."—E. J. Tangerman. 118pp. 7¾ × 10⅝. 23654-4 Pa. $5.95

ILLUSTRATED CATALOG OF CIVIL WAR MILITARY GOODS: Union Army Weapons, Insignia, Uniform Accessories, and Other Equipment, Schuyler, Hartley, and Graham. Rare, profusely illustrated 1846 catalog includes Union Army uniform and dress regulations, arms and ammunition, coats, insignia, flags, swords, rifles, etc. 226 illustrations. 160pp. 9 × 12. 24939-5 Pa. $10.95

WOMEN'S FASHIONS OF THE EARLY 1900s: An Unabridged Republication of "New York Fashions, 1909," National Cloak & Suit Co. Rare catalog of mail-order fashions documents women's and children's clothing styles shortly after the turn of the century. Captions offer full descriptions, prices. Invaluable resource for fashion, costume historians. Approximately 725 illustrations. 128pp. 8⅜ × 11¼. 27276-1 Pa. $11.95

THE 1912 AND 1915 GUSTAV STICKLEY FURNITURE CATALOGS, Gustav Stickley. With over 200 detailed illustrations and descriptions, these two catalogs are essential reading and reference materials and identification guides for Stickley furniture. Captions cite materials, dimensions and prices. 112pp. 6½ × 9¼. 26676-1 Pa. $9.95

EARLY AMERICAN LOCOMOTIVES, John H. White, Jr. Finest locomotive engravings from early 19th century: historical (1804–74), main-line (after 1870), special, foreign, etc. 147 plates. 142pp. 11⅜ × 8¼. 22772-3 Pa. $8.95

THE TALL SHIPS OF TODAY IN PHOTOGRAPHS, Frank O. Braynard. Lavishly illustrated tribute to nearly 100 majestic contemporary sailing vessels: Amerigo Vespucci, Clearwater, Constitution, Eagle, Mayflower, Sea Cloud, Victory, many more. Authoritative captions provide statistics, background on each ship. 190 black-and-white photographs and illustrations. Introduction. 128pp. 8⅞ × 11¾. 27163-3 Pa. $13.95

EARLY NINETEENTH-CENTURY CRAFTS AND TRADES, Peter Stockham (ed.). Extremely rare 1807 volume describes to youngsters the crafts and trades of the day: brickmaker, weaver, dressmaker, bookbinder, ropemaker, saddler, many more. Quaint prose, charming illustrations for each craft. 20 black-and-white line illustrations. 192pp. 4⅝ × 6. 27293-1 Pa. $4.95

VICTORIAN FASHIONS AND COSTUMES FROM HARPER'S BAZAR, 1867–1898, Stella Blum (ed.). Day costumes, evening wear, sports clothes, shoes, hats, other accessories in over 1,000 detailed engravings. 320pp. 9⅜ × 12¼. 22990-4 Pa. $13.95

GUSTAV STICKLEY, THE CRAFTSMAN, Mary Ann Smith. Superb study surveys broad scope of Stickley's achievement, especially in architecture. Design philosophy, rise and fall of the Craftsman empire, descriptions and floor plans for many Craftsman houses, more. 86 black-and-white halftones. 31 line illustrations. Introduction. 208pp. 6½ × 9¼. 27210-9 Pa. $9.95

THE LONG ISLAND RAIL ROAD IN EARLY PHOTOGRAPHS, Ron Ziel. Over 220 rare photos, informative text document origin (1844) and development of rail service on Long Island. Vintage views of early trains, locomotives, stations, passengers, crews, much more. Captions. 8⅞ × 11¾. 26301-0 Pa. $13.95

THE BOOK OF OLD SHIPS: From Egyptian Galleys to Clipper Ships, Henry B. Culver. Superb, authoritative history of sailing vessels, with 80 magnificent line illustrations. Galley, bark, caravel, longship, whaler, many more. Detailed, informative text on each vessel by noted naval historian. Introduction. 256pp. 5⅜ × 8½. 27332-6 Pa. $6.95

TEN BOOKS ON ARCHITECTURE, Vitruvius. The most important book ever written on architecture. Early Roman aesthetics, technology, classical orders, site selection, all other aspects. Morgan translation. 331pp. 5⅜ × 8½. 20645-9 Pa. $8.95

THE HUMAN FIGURE IN MOTION, Eadweard Muybridge. More than 4,500 stopped-action photos, in action series, showing undraped men, women, children jumping, lying down, throwing, sitting, wrestling, carrying, etc. 390pp. 7⅞ × 10⅝. 20204-6 Clothbd. $24.95

TREES OF THE EASTERN AND CENTRAL UNITED STATES AND CANADA, William M. Harlow. Best one-volume guide to 140 trees. Full descriptions, woodlore, range, etc. Over 600 illustrations. Handy size. 288pp. 4½ × 6⅜. 20395-6 Pa. $5.95

SONGS OF WESTERN BIRDS, Dr. Donald J. Borror. Complete song and call repertoire of 60 western species, including flycatchers, juncoes, cactus wrens, many more—includes fully illustrated booklet. Cassette and manual 99913-0 $8.95

GROWING AND USING HERBS AND SPICES, Milo Miloradovich. Versatile handbook provides all the information needed for cultivation and use of all the herbs and spices available in North America. 4 illustrations. Index. Glossary. 236pp. 5⅜ × 8½. 25058-X Pa. $5.95

BIG BOOK OF MAZES AND LABYRINTHS, Walter Shepherd. 50 mazes and labyrinths in all—classical, solid, ripple, and more—in one great volume. Perfect inexpensive puzzler for clever youngsters. Full solutions. 112pp. 8⅛ × 11. 22951-3 Pa. $3.95

PIANO TUNING, J. Cree Fischer. Clearest, best book for beginner, amateur. Simple repairs, raising dropped notes, tuning by easy method of flattened fifths. No previous skills needed. 4 illustrations. 201pp. 5⅜ × 8½.   23267-0 Pa. $5.95

A SOURCE BOOK IN THEATRICAL HISTORY, A. M. Nagler. Contemporary observers on acting, directing, make-up, costuming, stage props, machinery, scene design, from Ancient Greece to Chekhov. 611pp. 5⅜ × 8½.   20515-0 Pa. $11.95

THE COMPLETE NONSENSE OF EDWARD LEAR, Edward Lear. All nonsense limericks, zany alphabets, Owl and Pussycat, songs, nonsense botany, etc., illustrated by Lear. Total of 320pp. 5⅜ × 8½. (USO)   20167-8 Pa. $6.95

VICTORIAN PARLOUR POETRY: An Annotated Anthology, Michael R. Turner. 117 gems by Longfellow, Tennyson, Browning, many lesser-known poets. "The Village Blacksmith," "Curfew Must Not Ring Tonight," "Only a Baby Small," dozens more, often difficult to find elsewhere. Index of poets, titles, first lines. xxiii + 325pp. 5⅜ × 8¼.   27044-0 Pa. $8.95

DUBLINERS, James Joyce. Fifteen stories offer vivid, tightly focused observations of the lives of Dublin's poorer classes. At least one, "The Dead," is considered a masterpiece. Reprinted complete and unabridged from standard edition. 160pp. 5³⁄₁₆ × 8¼.   26870-5 Pa. $1.00

THE HAUNTED MONASTERY and THE CHINESE MAZE MURDERS, Robert van Gulik. Two full novels by van Gulik, set in 7th-century China, continue adventures of Judge Dee and his companions. An evil Taoist monastery, seemingly supernatural events; overgrown topiary maze hides strange crimes. 27 illustrations. 328pp. 5⅜ × 8½.   23502-5 Pa. $7.95

THE BOOK OF THE SACRED MAGIC OF ABRAMELIN THE MAGE, translated by S. MacGregor Mathers. Medieval manuscript of ceremonial magic. Basic document in Aleister Crowley, Golden Dawn groups. 268pp. 5⅜ × 8½.
23211-5 Pa. $8.95

NEW RUSSIAN-ENGLISH AND ENGLISH-RUSSIAN DICTIONARY, M. A. O'Brien. This is a remarkably handy Russian dictionary, containing a surprising amount of information, including over 70,000 entries. 366pp. 4½ × 6⅛.
20208-9 Pa. $9.95

HISTORIC HOMES OF THE AMERICAN PRESIDENTS, Second, Revised Edition, Irvin Haas. A traveler's guide to American Presidential homes, most open to the public, depicting and describing homes occupied by every American President from George Washington to George Bush. With visiting hours, admission charges, travel routes. 175 photographs. Index. 160pp. 8¼ × 11. 26751-2 Pa. $10.95

NEW YORK IN THE FORTIES, Andreas Feininger. 162 brilliant photographs by the well-known photographer, formerly with *Life* magazine. Commuters, shoppers, Times Square at night, much else from city at its peak. Captions by John von Hartz. 181pp. 9¼ × 10¾.   23585-8 Pa. $12.95

INDIAN SIGN LANGUAGE, William Tomkins. Over 525 signs developed by Sioux and other tribes. Written instructions and diagrams. Also 290 pictographs. 111pp. 6⅛ × 9¼.   22029-X Pa. $3.50

ANATOMY: A Complete Guide for Artists, Joseph Sheppard. A master of figure drawing shows artists how to render human anatomy convincingly. Over 460 illustrations. 224pp. 8⅜ × 11¼. 27279-6 Pa. $9.95

MEDIEVAL CALLIGRAPHY: Its History and Technique, Marc Drogin. Spirited history, comprehensive instruction manual covers 13 styles (ca. 4th century thru 15th). Excellent photographs; directions for duplicating medieval techniques with modern tools. 224pp. 8⅜ × 11¼. 26142-5 Pa. $11.95

DRIED FLOWERS: How to Prepare Them, Sarah Whitlock and Martha Rankin. Complete instructions on how to use silica gel, meal and borax, perlite aggregate, sand and borax, glycerine and water to create attractive permanent flower arrangements. 12 illustrations. 32pp. 5⅜ × 8½. 21802-3 Pa. $1.00

EASY-TO-MAKE BIRD FEEDERS FOR WOODWORKERS, Scott D. Campbell. Detailed, simple-to-use guide for designing, constructing, caring for and using feeders. Text, illustrations for 12 classic and contemporary designs. 96pp. 5⅜ × 8½. 25847-5 Pa. $2.95

OLD-TIME CRAFTS AND TRADES, Peter Stockham. An 1807 book created to teach children about crafts and trades open to them as future careers. It describes in detailed, nontechnical terms 24 different occupations, among them coachmaker, gardener, hairdresser, lacemaker, shoemaker, wheelwright, copper-plate printer, milliner, trunkmaker, merchant and brewer. Finely detailed engravings illustrate each occupation. 192pp. 4⅝ × 6. 27398-9 Pa. $4.95

THE HISTORY OF UNDERCLOTHES, C. Willett Cunnington and Phyllis Cunnington. Fascinating, well-documented survey covering six centuries of English undergarments, enhanced with over 100 illustrations: 12th-century laced-up bodice, footed long drawers (1795), 19th-century bustles, 19th-century corsets for men, Victorian "bust improvers," much more. 272pp. 5⅜ × 8¼. 27124-2 Pa. $9.95

ARTS AND CRAFTS FURNITURE: The Complete Brooks Catalog of 1912, Brooks Manufacturing Co. Photos and detailed descriptions of more than 150 now very collectible furniture designs from the Arts and Crafts movement depict davenports, settees, buffets, desks, tables, chairs, bedsteads, dressers and more, all built of solid, quarter-sawed oak. Invaluable for students and enthusiasts of antiques, Americana and the decorative arts. 80pp. 6½ × 9¼. 27471-3 Pa. $7.95

HOW WE INVENTED THE AIRPLANE: An Illustrated History, Orville Wright. Fascinating firsthand account covers early experiments, construction of planes and motors, first flights, much more. Introduction and commentary by Fred C. Kelly. 76 photographs. 96pp. 8¼ × 11. 25662-6 Pa. $8.95

THE ARTS OF THE SAILOR: Knotting, Splicing and Ropework, Hervey Garrett Smith. Indispensable shipboard reference covers tools, basic knots and useful hitches; handsewing and canvas work, more. Over 100 illustrations. Delightful reading for sea lovers. 256pp. 5⅜ × 8½. 26440-8 Pa. $7.95

FRANK LLOYD WRIGHT'S FALLINGWATER: The House and Its History, Second, Revised Edition, Donald Hoffmann. A total revision—both in text and illustrations—of the standard document on Fallingwater, the boldest, most personal architectural statement of Wright's mature years, updated with valuable new material from the recently opened Frank Lloyd Wright Archives. "Fascinating"—The New York Times. 116 illustrations. 128pp. 9¼ × 10¾. 27430-6 Pa. $10.95

THE FOUR-COLOR PROBLEM: Assaults and Conquest, Thomas L. Saaty and Paul G. Kainen. Engrossing, comprehensive account of the century-old combinatorial topological problem, its history and solution. Bibliographies. Index. 110 figures. 228pp. 5⅜ × 8½. 65092-8 Pa. $6.95

CATALYSIS IN CHEMISTRY AND ENZYMOLOGY, William P. Jencks. Exceptionally clear coverage of mechanisms for catalysis, forces in aqueous solution, carbonyl- and acyl-group reactions, practical kinetics, more. 864pp. 5⅜ × 8½. 65460-5 Pa. $19.95

PROBABILITY: An Introduction, Samuel Goldberg. Excellent basic text covers set theory, probability theory for finite sample spaces, binomial theorem, much more. 360 problems. Bibliographies. 322pp. 5⅜ × 8½. 65252-1 Pa. $8.95

LIGHTNING, Martin A. Uman. Revised, updated edition of classic work on the physics of lightning. Phenomena, terminology, measurement, photography, spectroscopy, thunder, more. Reviews recent research. Bibliography. Indices. 320pp. 5⅜ × 8¼. 64575-4 Pa. $8.95

PROBABILITY THEORY: A Concise Course, Y.A. Rozanov. Highly readable, self-contained introduction covers combination of events, dependent events, Bernoulli trials, etc. Translation by Richard Silverman. 148pp. 5⅜ × 8¼.
63544-9 Pa. $5.95

AN INTRODUCTION TO HAMILTONIAN OPTICS, H. A. Buchdahl. Detailed account of the Hamiltonian treatment of aberration theory in geometrical optics. Many classes of optical systems defined in terms of the symmetries they possess. Problems with detailed solutions. 1970 edition. xv + 360pp. 5⅜ × 8½.
67597-1 Pa. $10.95

STATISTICS MANUAL, Edwin L. Crow, et al. Comprehensive, practical collection of classical and modern methods prepared by U.S. Naval Ordnance Test Station. Stress on use. Basics of statistics assumed. 288pp. 5⅜ × 8½.
60599-X Pa. $6.95

DICTIONARY/OUTLINE OF BASIC STATISTICS, John E. Freund and Frank J. Williams. A clear concise dictionary of over 1,000 statistical terms and an outline of statistical formulas covering probability, nonparametric tests, much more. 208pp. 5⅜ × 8½. 66796-0 Pa. $6.95

STATISTICAL METHOD FROM THE VIEWPOINT OF QUALITY CONTROL, Walter A. Shewhart. Important text explains regulation of variables, uses of statistical control to achieve quality control in industry, agriculture, other areas. 192pp. 5⅜ × 8½. 65232-7 Pa. $7.95

THE INTERPRETATION OF GEOLOGICAL PHASE DIAGRAMS, Ernest G. Ehlers. Clear, concise text emphasizes diagrams of systems under fluid or containing pressure; also coverage of complex binary systems, hydrothermal melting, more. 288pp. 6½ × 9¼. 65389-7 Pa. $10.95

STATISTICAL ADJUSTMENT OF DATA, W. Edwards Deming. Introduction to basic concepts of statistics, curve fitting, least squares solution, conditions without parameter, conditions containing parameters. 26 exercises worked out. 271pp. 5⅜ × 8½. 64685-8 Pa. $8.95

TENSOR CALCULUS, J.L. Synge and A. Schild. Widely used introductory text covers spaces and tensors, basic operations in Riemannian space, non-Riemannian spaces, etc. 324pp. 5⅜ × 8¼. 63612-7 Pa. $8.95

A CONCISE HISTORY OF MATHEMATICS, Dirk J. Struik. The best brief history of mathematics. Stresses origins and covers every major figure from ancient Near East to 19th century. 41 illustrations. 195pp. 5⅜ × 8½. 60255-9 Pa. $7.95

A SHORT ACCOUNT OF THE HISTORY OF MATHEMATICS, W.W. Rouse Ball. One of clearest, most authoritative surveys from the Egyptians and Phoenicians through 19th-century figures such as Grassman, Galois, Riemann. Fourth edition. 522pp. 5⅜ × 8½. 20630-0 Pa. $10.95

HISTORY OF MATHEMATICS, David E. Smith. Nontechnical survey from ancient Greece and Orient to late 19th century; evolution of arithmetic, geometry, trigonometry, calculating devices, algebra, the calculus. 362 illustrations. 1,355pp. 5⅜ × 8½. 20429-4, 20430-8 Pa., Two-vol. set $23.90

THE GEOMETRY OF RENÉ DESCARTES, René Descartes. The great work founded analytical geometry. Original French text, Descartes' own diagrams, together with definitive Smith-Latham translation. 244pp. 5⅜ × 8½.
60068-8 Pa. $7.95

THE ORIGINS OF THE INFINITESIMAL CALCULUS, Margaret E. Baron. Only fully detailed and documented account of crucial discipline: origins; development by Galileo, Kepler, Cavalieri; contributions of Newton, Leibniz, more. 304pp. 5⅜ × 8½. (Available in U.S. and Canada only) 65371-4 Pa. $9.95

THE HISTORY OF THE CALCULUS AND ITS CONCEPTUAL DEVELOPMENT, Carl B. Boyer. Origins in antiquity, medieval contributions, work of Newton, Leibniz, rigorous formulation. Treatment is verbal. 346pp. 5⅜ × 8½.
60509-4 Pa. $8.95

THE THIRTEEN BOOKS OF EUCLID'S ELEMENTS, translated with introduction and commentary by Sir Thomas L. Heath. Definitive edition. Textual and linguistic notes, mathematical analysis. 2,500 years of critical commentary. Not abridged. 1,414pp. 5⅜ × 8½. 60088-2, 60089-0, 60090-4 Pa., Three-vol. set $29.85

GAMES AND DECISIONS: Introduction and Critical Survey, R. Duncan Luce and Howard Raiffa. Superb nontechnical introduction to game theory, primarily applied to social sciences. Utility theory, zero-sum games, n-person games, decision-making, much more. Bibliography. 509pp. 5⅜ × 8½. 65943-7 Pa. $12.95

THE HISTORICAL ROOTS OF ELEMENTARY MATHEMATICS, Lucas N.H. Bunt, Phillip S. Jones, and Jack D. Bedient. Fundamental underpinnings of modern arithmetic, algebra, geometry and number systems derived from ancient civilizations. 320pp. 5⅜ × 8½. 25563-8 Pa. $8.95

CALCULUS REFRESHER FOR TECHNICAL PEOPLE, A. Albert Klaf. Covers important aspects of integral and differential calculus via 756 questions. 566 problems, most answered. 431pp. 5⅜ × 8½. 20370-0 Pa. $8.95

CATALOG OF DOVER BOOKS

CHALLENGING MATHEMATICAL PROBLEMS WITH ELEMENTARY SOLUTIONS, A.M. Yaglom and I.M. Yaglom. Over 170 challenging problems on probability theory, combinatorial analysis, points and lines, topology, convex polygons, many other topics. Solutions. Total of 445pp. 5⅜ × 8½. Two-vol. set.

Vol. I 65536-9 Pa. $7.95

Vol. II 65537-7 Pa. $6.95

FIFTY CHALLENGING PROBLEMS IN PROBABILITY WITH SOLUTIONS, Frederick Mosteller. Remarkable puzzlers, graded in difficulty, illustrate elementary and advanced aspects of probability. Detailed solutions. 88pp. 5⅜ × 8½.

65355-2 Pa. $4.95

EXPERIMENTS IN TOPOLOGY, Stephen Barr. Classic, lively explanation of one of the byways of mathematics. Klein bottles, Moebius strips, projective planes, map coloring, problem of the Koenigsberg bridges, much more, described with clarity and wit. 43 figures. 210pp. 5⅜ × 8½.

25933-1 Pa. $5.95

RELATIVITY IN ILLUSTRATIONS, Jacob T. Schwartz. Clear nontechnical treatment makes relativity more accessible than ever before. Over 60 drawings illustrate concepts more clearly than text alone. Only high school geometry needed. Bibliography. 128pp. 6⅛ × 9¼.

25965-X Pa. $6.95

AN INTRODUCTION TO ORDINARY DIFFERENTIAL EQUATIONS, Earl A. Coddington. A thorough and systematic first course in elementary differential equations for undergraduates in mathematics and science, with many exercises and problems (with answers). Index. 304pp. 5⅜ × 8½.

65942-9 Pa. $8.95

FOURIER SERIES AND ORTHOGONAL FUNCTIONS, Harry F. Davis. An incisive text combining theory and practical example to introduce Fourier series, orthogonal functions and applications of the Fourier method to boundary-value problems. 570 exercises. Answers and notes. 416pp. 5⅜ × 8½.

65973-9 Pa. $9.95

THE THEORY OF BRANCHING PROCESSES, Theodore E. Harris. First systematic, comprehensive treatment of branching (i.e. multiplicative) processes and their applications. Galton-Watson model, Markov branching processes, electron-photon cascade, many other topics. Rigorous proofs. Bibliography. 240pp. 5⅜ × 8½.

65952-6 Pa. $6.95

AN INTRODUCTION TO ALGEBRAIC STRUCTURES, Joseph Landin. Superb self-contained text covers "abstract algebra": sets and numbers, theory of groups, theory of rings, much more. Numerous well-chosen examples, exercises. 247pp. 5⅜ × 8½.

65940-2 Pa. $7.95
